Traces of Ink

Nuncius Series

Studies and Sources in the Material and Visual History of Science

Series Editors

Marco Beretta (*University of Bologna*)
Sven Dupré (*Utrecht University / University of Amsterdam*)

VOLUME 7

The titles published in this series are listed at *brill.com/nuns*

Traces of Ink

Experiences of Philology and Replication

Edited by

Lucia Raggetti

BRILL

LEIDEN | BOSTON

This volume is published within the framework of the ERC Project AlchemEast — *Alchemy in the Making: From ancient Babylonia via Graeco-Roman Egypt into the Byzantine, Syriac and Arabic traditions* (*1500 BCE–1000 AD*), which has received funding from the European Research Council (ERC) under the European Union's Horizon 2020 research and innovation programme (G.A. 724914).

Cover illustration: Composition of writing implements. © Photo by Lucia Raggetti 2020, from objects in the Ter Lugt Collection, Leiden. In the foreground: a Byzantine bronze stylus, a modern reed pen (*qalam*), bookbinder's scissors showing the text *Yā Fattāḥ* (Ottoman Egypt), ceramic figurine of a Muslim scholar (Samarkand, late 20th-century), wooden penholder from Northern India. In the background: a carved stone inkwell (Central Asia), a glazed ceramic inkwells (Central Asia), and a bone paper burnisher (Ottoman Egypt).

The Library of Congress Cataloging-in-Publication Data is available online at http://catalog.loc.gov
LC record available at http://lccn.loc.gov/2020054333

Typeface for the Latin, Greek, and Cyrillic scripts: "Brill". See and download: brill.com/brill-typeface.

ISSN 2405-5077
ISBN 978-90-04-42111-0 (hardback)
ISBN 978-90-04-44480-5 (e-book)

This book is printed on acid-free paper and produced in a sustainable manner.

Contents

Acknowledgements

I would like to thank the other members of the *AlchemEast* team, in particular Miriam Blanco Cesteros and Matteo Martelli, for their constant support: I could count on their precious and learned advice in all the preparatory stages of this book, tight spots included. For the linguistic revision of the introduction, my thanks go to my friends and colleagues Sean Coughlin and Peter Singer. I am very grateful to Sean Coughlin also for all the enthusiastic and inspiring discussions about replication.

Figures

Notes on Contributors

Miriam Blanco Cesteros
is currently a postdoctoral fellow at the University Complutense of Madrid (Spain). With a solid background in papyrology and classical philology, she received her PhD with a dissertation on Greco-Egyptian magical papyri in 2017. In 2018–2020 she was a postdoctoral fellow in the ERC project *AlchemEast* on the history of Alchemy (Bologna, Italy), where she developed a research focused on the two Greco-Egyptian alchemical recipe books usually referred to as the Leiden and Stockholm papyri (3rd–4th century AD), to be published as a monograph. Her research revolves around the ritual and cultural formation of the Greco-Egyptian magic practitioners, and her publications touch upon a variety of fields such as magic and religion, literature, alchemy and papyrology in an interdisciplinary perspective.

Michele Cammarosano
(post-doc, Philipps-Universität Marburg) was awarded a PhD in Ancient Near Eastern Studies from the University of Naples "L'Orientale" in 2012. His research focuses on Hittite religion and administration as well as on the investigation of writing techniques, with a strong interest in the exploration of interdisciplinary approaches involving the combination of archaeology, philology, and informatics. His last book is *Hittite Local Cults* (Atlanta 2018).

Claudia Colini
is a researcher at the Federal Institute for Materials Research and Testing (BAM) and a member of the Cluster of Excellence "Understanding written artefacts," University of Hamburg. Her current project is about the coexistence of different writing supports and inks in Egypt in the early centuries of Islam. In 2018 she completed her doctorate at the CSMC, University of Hamburg, Germany, with a dissertation entitled "From recipes to material analysis: the Arabic tradition of black inks and paper coatings (9th–20th century)," which she is currently revising for publication as a monograph. Her dissertation and a number of related publications focus on the identification and history of materials and techniques used in the production of Islamic manuscripts.

Vincenzo Damiani
received his PhD at the Scuola Normale Superiore di Pisa and the Julius-Maximilians-Universität Würzburg with a Dissertation on didactic compendia in Epicureanism (2019), which is in publication. Formerly a research associate

at the Institute of Classics of the Julius-Maximilians-Universität Würzburg, he is currently working as an assistant lecturer at the Institute of the History, Philosophy and Ethics of Medicine, Ulm University. He has published on Herculaneum Papyrology (editio princeps of PHerc. 1026), Epicurean philosophy, and ancient science, with a particular emphasis on ancient medicine.

Sara Fani

is currently post-Doc fellow at the University of Naples "L'Orientale" for the project *The European Qur'an*. She has previously been Adjunct professor of Arabic Literature and post-doc fellow at the University of Florence (2019) and post-Doc fellow at the University of Copenhagen (2014–2018). In 2017 she collaborated with the Kunsthistorisches Institut in Florenz for the project *Die Typographia Medicea im Kontext*. She has combined her literary interest in Arabic technical texts of codicological subject with a Master's degree in conservation of book materials and with the study of manuscript collections. Her present research focuses on early-Modern Italian approach to Islam.

Matteo Martelli

(PhD Greek Philology, 2007; PhD History of Science, 2012) is professor of History of Science at the University of Bologna. His research focuses on Graeco-Roman and Byzantine science — with particular attention to alchemy and medicine — and its reception in the Syro-Arabic tradition. His publications include *The Four Books of Pseudo-Democritus* (2014) and *Collecting Recipes. Byzantine and Jewish Pharmacology in Dialogue* (2017; edited with L. Lehmhaus). He is the principal investigator of the ERC project *AlchemEast*, and he is currently working on a critical edition of the Syriac alchemical books ascribed to Zosimos of Panopolis.

Ira Rabin

is professor at the Hamburg University and scientist at the Federal Institute of Material Research and Testing (BAM) in Berlin. Until 2003 she worked in basic research in cluster physics in the Fritz-Haber-Institute of the Max-Planck-Society in Berlin. Since 2003 her main research interest has been dedicated to the Dead Sea Scrolls. Currently besides conducting research dedicated to re-construction of the history of black writing inks she is working towards including ink composition into codicological studies.

Lucia Raggetti

is an Assistant Professor for the History of Ancient Sciences at the University of Bologna. After receiving her PhD in Arabo-Islamic studies in Naples, she held a DAAD Fellowship in Hamburg and then worked as research assistant at the Freie Unversität Berlin, in the research group on *Wissensgeschichte* (History of Knowledge). Her main research interests are Arabic philology and the history of natural sciences and medicine in the Arabo-Islamic milieu, on which she has published a variety of articles. She is author of *ʿĪsā ibn ʿAlī's Book on the Useful Properties of Animal Parts: Edition, Translation and Study of a Fluid Tradition* (Berlin 2018).

Katja Weirauch
is lecturer for Chemistry teaching at the University of Würzburg, Germany. Previously she taught Biology and Chemistry in various German regions and high schools (2000–2010) and was the editor for Chemistry at the German governmental institution for educational media (FWU Medieninstitut der Bundesländer, 2004–2007). Her current research interests are emotionally effective and authentic contexts for Chemistry teaching (e.g. historical ones) and the development of and empirical research about settings for inclusive Science teaching. Her latest book *Chemie all-inclusive* discusses how to design inclusive experiments for heterogenous learning groups (Würzburg 2020).

Introduction

Lucia Raggetti

ربع الكتابة من سواد مدادها   والربع حسن صناعة الكتّاب
والربع من قلم تسوي بريه   وعلى الكواغد رابع الاسباب

One-fourth of writing derives from its ink's blackness,
from a beautifully made book another fourth,
One-fourth from a well sharpened pen,
while on the leaves of paper rests the fourth cause.
AL-QALQAŠANDĪ, *Subḥ al-aʿšā* II, 502

∴

An Arabic couplet, which circulated in Mediaeval erudite and scribal circles, says that the art of penmanship results from the sum of four different elements: the blackness of the ink, the making of a beautiful written artefact, a sharp pen, and a good sheet of paper. This aphoristic hemistich singles out the pivotal components common to almost all acts of writing: the presence of a medium, a suitable support, along with the use of ergonomic writing implements in order to obtain a final result that has aesthetic value. This synthetic perspective allows one to approach many different cultures and historical periods together, without losing sight of the differences between their particular interpretations of the shared pattern in writing practices. Indeed, premodern cultures reflected on writing supports and implements, as well as the composition of written artefacts, and on the tools and substances which were meant to leave a clear impression of written signs. They left behind a variety of material and textual witnesses, all of which were influenced and shaped by the language they spoke, the writing system they adopted, and the cultural lore related to the act of writing and its social meaning.

Recently, scholarship has developed a keen interest in following the 'traces of ink' left by many different premodern cultures. To mention but a few of these titles here would not do justice to the scholarly fields and specializations involved in the exploration of this research field. I leave it to the chapters in this volume to offer a rich and up-to-date bibliography on their different subjects.

Scholarly interest in inks tends to branch out in two opposite directions: one branch directed towards the study of the material aspects of writing; the other towards the textual sources dealing with the act and art of writing itself. There are several disciplines engaged in the study of the material aspects (codicology, conservation, archaeology, chemistry), while philology usually considers the texts. This distinction, however, is rather artificial and definitely influenced by the boundaries of modern academic disciplines. The material and the textual aspect, in fact, do not exist in monadic isolation: there are large areas of intersection between the two, spaces in which these two components are indissolubly intertwined, though one may occasionally appear predominant over the other. This scenario calls for an interdisciplinary approach that involves both concrete cooperation between scholars of different disciplines and a new mindset and research perspective that individual researchers may adopt.

To start with the 'textual' component, philology has sometimes looked to the text as an ideal and metaphysical entity, one which exists detached and disjoined from any kind of materiality. Moreover, the editorial work on technical texts — and collections of ink recipes and handbook on penmanship represent one possible example — should take into account that technical literature has its peculiar traits. Recipes are textual units characterized by a great syntactic coherence and, at the same time, an inclination to attract textual variants of a different nature, which altogether poses specific philological challenges. Recipes are, in fact, textual blocks that can circulate quite freely and fluidly, finding ways to fit into new textual structures and collections. Copying recipes is not only a mechanical process and a fatal occasion to introduce mistakes, more often it represents a chance to update, refine and personalise the text. In their wandering transmission, these textual units show no inclination to remain confined within a single genre, which calls for a very inclusive perusal of the sources. The attempt to give a philological account of recipes cannot focus on the obsessive research into an alleged 'archetype' or original written version recording the genuine form of a certain procedure. This archetype is very likely to be out of reach. The aim, rather, is to understand the genesis of meaningful variants, at the same time trying to reconstruct the movements of these textual blocks not only from one collection to the other, but also between different linguistic and cultural contexts. Neo-Lachmannian philology — as developed in the second half of the twentieth century by Giorgio Pasquali, Michele Barbi and Gianfranco Contini, to name just a few of the 'giants' in the field — has always paid attention to the material and cultural dimension of the manuscript witness and to the fluidity of the transmission. Recent publications, such as the COMSt Handbook (*Comparative Oriental Manuscript Studies: An Introduction*, 2015) and Paolo Trovato's *Everything you Always Wanted to*

Know about Lachmann's Method. A non-standard handbook of genealogical textual criticism in the age of post-structuralism, cladistics (2017), have allowed these ideas to expand their circulation beyond the relatively small circles of those who read Italian.

As for the material aspects, when approaching the study of premodern technical and scientific practices, contemporary scientists are bound to consider that premodern technical knowledge was conceptualized and expressed in its own peculiar way, and that some of its components are deeply influenced by the different literary traditions and by the transmission of the text itself. While carrying out scientific analysis of inks, it is inadvisable to lose sight of the fact that a trace of ink is not a spot occasionally produced on a random surface long ago, but originates from a purposeful act of writing with the intention of conveying meaning.

When the experts from the two sides — the textual and the technical — inform each other, they can create a virtuous circle that promotes the conditions for a larger and deeper understating of premodern literary and technical traditions.

Replication, in particular, represents a privileged experience in this field of research, opened up by Lawrence Principe's pioneering work. In concrete terms, replication can be defined as the attempt to reproduce the material and chemical reality behind a text, to better understand the relation between its contents and the written form in which these were recorded. In this way, replication becomes much more than an empirical way to assess the success rate of a certain procedure, or the lack thereof. Replication is rather meant to shed light on the identification of the ingredients, to provide a concrete referent corresponding to some flowery and cryptic description of the processes involved, to sift the technical contents from the elements defined by literary tradition and manuscript transmission that is, in other words, to grasp the genesis and the character of variants. The well-balanced integration between the textual and the material dimension opens up a new interdisciplinary research space, in which each expert is invited to share the best of his knowledge in a way that may be intelligible outside the restricted circle of fellow specialists. An interdisciplinary approach combined with replication, as they have just been described, also represents a methodological safety-net, which protects us from falling into circular arguments in which a unilateral hypothesis and the results of a procedure turn into their own proof.

Every chapter in this volume deals with a specific case study set in the area of intersection between the material and textual aspects of inks in different premodern written traditions. Apart from Chapters 3 and 5 — included to enrich the historical overview of the volume — the chapters of this book derive from papers originally presented and discussed during the workshop *Traces of Inks* (12th July 2018, Bologna, organized in the frame of the ERC Project *AlchemEast*) that gathered scholars from different academic fields working on inks and written artefacts.

The chapters of this volume are arranged in chronological order and propose a long durée perspective, that starts from Babylonia and the Ancient Near East (Chapter 1), has the 13th-century Arabo-Islamic tradition as its farthest chronological point (Chapters 6–8), with a number of relevant cases from antiquity and late antiquity in between (Chapters 2–5). There are, however, thematic threads that connect the different chapters beyond their position on the time line of history. Such threads are not loosely juxtaposed: they represent the warp and weft of the interdisciplinary space of research explored in this volume.

In Chapter 4 (*Material Studies of Historic Inks: Transition from Carbon to Iron-Gall Inks*), the exploration of the material aspects finds a large historical overview of black inks from antiquity to the Middle Ages, along with the modern methods of analysis employed for their detection and identification. A similar line is traced by other chapters that focus on specific writing traditions. Chapter 3 (*Ink in Herculaneum: A Survey of Recent Perspectives*) is a close-up on the case of the Herculaneum papyri, in which the story of these unique specimens intertwines with the technological developments in their study. Chapter 7 (*"I tried it and it is really good" Replicating Recipes of Arabic Black Inks*) scrutinizes the technical contribution to philology, the feasibility of the procedures, and what these reveal about the technical skills of compilers from the point of view of a book conservator working on Arabic ink recipes and their replication, discussing the possible role of replicated inks as reference for scientific analysis. Chapter 1 (*WoW! Writing on Wax in Ancient Mesopotamia and Today: Questions and Results from an Interdisciplinary Project*) is the result of the collaboration between a philologist and a chemist, working together on ancient wax boards, and exploring current paedagogical application of their research.

The philological and textual thread emerges in connection with the literary nature of sources dealing with inks and the original context and intent of their composition, exploring interpretative and editorial possibilities in a number of different cultural and linguistic traditions. Chapter 3 (*Written in Blood? Decoding Some Red Inks of the Greek Magical Papyri*), for instance, examines the occurrences of blood as ingredient for inks in the Greek magical papyri in order to unveil its status as code name and the effects of this phenomenon on the textual transmission. The Syriac ink recipes, transmitted in alchemical sources as discussed in Chapter 5 (*'Alchemical' Inks in the Syriac Tradition*) offer an example of the circulation of ink-related materials across the boundaries of different languages and textual genres, with special attention to the alchemical tradition. Inks in the Arabo-Islamic culture also provide the focus of other two chapters. Chapter 6 (*The Literary Dimension and Life of Arabic Treatises on Ink Making*) and Chapter 8 (*Ordinary Inks and Incredible Tricks in al-ʿIrāqī's ʿUyūn al-ḥaqāʾiq*) both explore the literary dimension of ink writings; the former focusing on the structure of technical treatises relating to ink-making and how this affected their transmission, while the latter concentrates on a 13th-century technical handbook on natural magic, leger-de-main and a variety of craft, proposing an editorial approach devised for interdisciplinary research on technical texts.

Apart from the value of collecting original and thought-provoking case studies, this volume aims to stress the importance of interdisciplinary research, showing the advantages of such a scholarly perspective and the interesting results that this attitude may bring to achieve. Inks and their traces are a productive field, a meaningful example and a fruitful opportunity to reflect upon the study of premodern science and technology and draw fresh methodological inspiration for future research.

CHAPTER 1

WoW! Writing on Wax in Ancient Mesopotamia and Today: Questions and Results from an Interdisciplinary Project

Katja Weirauch and Michele Cammarosano

Abstract

By dispensing with the need for ink, while simultaneously providing a writing surface that retains plasticity over time, for four millennia wax boards represented the precursors of modern "smart tablets," and are therefore one of the most relevant media in human history. They consist of one or more 'leaves' provided with a recessed frame that holds a beeswax-based mixture on which marks can be scratched or impressed. Today, wax boards are no longer used in everyday life; nevertheless, they provide new and unexpected opportunities for extracurricular learning. This chapter discusses the earliest history of wax boards, as attested in the cuneiform cultures of the Ancient Near East. It compares the boards from this period with those from Classical antiquity and the Middle Ages and subsequently focuses on a cross-disciplinary pedagogical concept for sixth grade classes. It integrates history and chemistry learning by involving the schoolchildren in the "making of science."

Keywords

wax boards – cuneiform writing – Ancient Near East – chemistry didactics – history didactics

1 On Wax, Tablets, and a "*Mission: Impossible*"

The earliest references to the use of wax boards date from the third millennium BCE. From a structural and functional point of view, wax diptychs can be truly considered as proto-codices, i.e. the earliest form of the modern book.[1]

1 See most recently Georgios Boudalis, *The Codex and Crafts in Late Antiquity* (New York: Bard Graduate Center, 2018), esp. p. 9 with fn. 21. The doubts expressed e.g. by John A. Szirmai,

DOI:10.1163/9789004444805_003

Used without interruption for more than three millennia for a variety of scripts in a number of different cultures, wax boards represent the longest-lived writing medium in human history. In the following discussion, we will focus on the earliest history of this medium as attested in the cuneiform cultures of the Ancient Near East, and on what those ancient times can teach us and today's schoolchildren. In doing so, we will embark on a fascinating journey at the intersection of philology, chemistry, and archaeology.

G. Boudalis made a number of observations in his brilliant study on the early history of the codex that are remarkably applicable to the reconstruction of the technology of ancient wax boards:

> The evidence for reconstructing the history of the Eastern Mediterranean tradition can be separated into three types: physical, iconographical, and literary. Inevitably, evidence for each is fragmentary, especially the first, which is our primary source of information; both iconographical and literary evidence, however, can at times be surprisingly accurate and helpful. [...] Iconographical evidence, if properly and consistently studied along with physical evidence, can help fill the many gaps related to the format and making of the book in late antiquity. [...] The main problem with exploiting iconographical evidence is that unless we have related physical evidence to which to compare it, we are often left with an image that we cannot quite decipher. Nevertheless, even when there are features we do not understand, they are often consistently represented in different works of art that are not connected in any apparent way, which suggests that they reflect the reality of the codex at the time, even if that reality remains unclear.[2]

Given the almost total absence of physical evidence, the lack of ancient treatises, and the role played by potentially dangerous substances, the investigation of the early technology of wax boards configures a sort of *mission impossible* that can only be addressed by a truly interdisciplinary approach.

The Archaeology of Medieval Bookbinding (Hampshire: Aldershot et al., 1999), p. 3 with fn. 1 seem to originate from a misinterpretation of Neo-Hittite reliefs, see the discussion in Michele Cammarosano, Katja Weirauch, Feline Maruhn, Gert Jendritzki, P. Kohl, "They Wrote on Wax. Wax Boards in the Ancient Near East," *Mesopotamia*, 2019, LIV:121–180, pp. 133–134 with fn. 126

2 Boudalis, *Codex and Crafts* (cit. note 1), pp. 15–16. We are indebted to Robert Fuchs for bringing this work to our attention. On the relevance of images for the reconstruction of material culture see, among others, Peter Burke, *Eyewitnessing. The uses of images as historical evidence* (London: Reaktion Books, 2001), pp. 81–102.

Interdisciplinarity is a fashionable word today, but it is easier said than done. For any researcher, keeping track of the developments within one's own field of expertise is a daily challenge, let alone going beyond it. That said, sometimes the most interesting insights arise precisely when distant disciplines meet. With its research foci in Biology and Chemistry as well as in Ancient Near Eastern Studies, the University of Würzburg provided ideal conditions for such an endeavour. This paper introduces some scientific and pedagogical products emerging from cooperation between the departments of Ancient Near Eastern Studies and Chemistry Teaching, and part of the results of the collaborative project "WoW! Writing on Wax."[3]

2 Writing on Wax

Wax boards are the forerunners to paper and the graphite pencil, which first appeared in the 16th century, and today's "smart tablets." From a functional perspective, these media share a fundamental characteristic: they all allow (1) writing without ink, and (2) the erasing and re-inscribing of written text *ad libitum*. In other words, they are all excellent technology for contexts requiring frequent correction of or addition to texts, especially if working outdoors. In the case of wax boards, this is achieved by using a stylus to scratch (or impress) marks in a layer of beeswax, most commonly mixed with a mineral pigment (and sometimes further additives) in order to optimize its mechanical and optical properties. The marks can be erased simply by passing a spatula or a

3 For a detailed discussion of the facts summarized in sections 2 to 5 of the present paper see Cammarosano et al., *They Wrote on Wax* (cit. note 1). It would not have been possible to carry out our project without the support and help of so many colleagues, students, and friends from various fields. We are particularly grateful to Robert Fuchs, Gert Jendritzki, Doris Oltrogge and Heinrich Piening for their invaluable help on a number of crucial issues related to the writing technology of wax boards and for sharing their unpublished research results with us. We are also indebted to Lutz Martin for granting us access to the collections of the *Vorderasiatisches Museum* in Berlin, to Astrid Nunn and Jürgen Tautz for help and advice in the earlier stages of the project, and to Jochen Griesbach, Geraldina Rozzi, Daniel Schwemer and Miron Sevastre for their contribution to the Cuneiform Lab in December 2016. Through her excellent thesis "Altorientalische Wachstafeln. Alte Medien neu erforscht" and her enthusiastic engagement in the project, Feline Maruhn has contributed much more to this chapter than is explicitly acknowledged here. We are very grateful to Matteo Martelli and Lucia Raggetti for the opportunity to present the project results at the workshop *Traces of Ink* and for the wonderful hospitality in Bologna in July 2018. Financial support from the *Universitätsbund Würzburg* (AZ 18–33) and from the chairs of Altorientalistik (Daniel Schwemer) and Didaktik der Chemie (Ekkehard Geidel) of the University of Würzburg is gratefully acknowledged.

globular tip over them, thereby allowing for immediate re-inscription of the surface. It is no coincidence, therefore, that, throughout history, wax boards and paper and pencil have been the privileged medium in the context of, in particular, school, bureaucracy, commerce, and the process of literary creation — precisely the contexts where tablet computers are most used today (fig. 1.1c).[4]

There is much evidence to illustrate such circumstances, some of which is exemplified in figures 1.1 and 1.2. The Greeks and Romans made extensive use of wax boards at school, as documented by textual and iconographic sources as well as by a great number of archaeological finds, mostly from Egypt. The Romans used wax boards not only for ephemeral records, but also for legal texts, such as contracts, meant to be preserved for a longer period. A remarkable number of such documents has been recovered in Pompeii and the surrounding area, mostly consisting of sealed, multi-page board books (fig. 1.2b–c).[5]

As mentioned, historically, wax boards have played a major role in the process of literary creation, as is immediately evident when one reads the poems of Propertius, Ovid and Martial, the *Institutio oratoria* by Quintilian, or Medieval anecdotal literature. Indeed, they allow the author to take notes when and where he wants — the depiction of the muse Calliope holding a wax diptych, from Triclinium A of the Inn of the Sulpicii in Murecine (fig. 1.2a), suggests that this was also true for supernatural beings.[6] An illumination from a manuscript dating ca. 1380 exemplifies the peculiar features of wax boards and

4 The image in fig. 1.1c is taken from an online advertisement for the Apple *iPad*.

5 Fig. 1.2b shows a reconstruction of one of the triptychs from the house of L. Caecilius Jucundus in Pompeii (from August Mau, *Pompeii. Its Life and Art* (New York: MacMillan, 1899), pp. 490–91), and G. Boudalis' reconstruction of a Roman polyptych based on a wall painting in Pompeii (from Boudalis, *Codex and Crafts* (cit. note 1), p. 25 fig. 11a, kindly provided by the author). On Greek and Roman wax boards, see Giuseppe Camodeca, "Gli archivi privati di tabulae ceratae e di papiri documentari a Pompei ed Ercolano: case, ambienti e modalità di conservazione," *Vesuviana: An international journal of archaeological and historical studies on Pompei and Herculaneum*, 2009, 1:17–42; Benjamin Hartmann, "Die hölzernen Schreibtafeln im Imperium Romanum — ein Inventar," in *Lesen und Schreiben in den Römischen Provinzen: Schriftliche Kommunikation im Alltagsleben*, edited by Markus Scholz, Marietta Horster (Mainz: Romisch-Germanisches Zentralmuseum, 2015), pp. 43–58; P. Tomlin, *Roman London's First Voices: Writing tablets from the Bloomberg excavations, 2010–14* (London: MOLA Museum of London Archeology, 2016); Paola Degni, *Usi delle tavolette lignee e cerate nel mondo greco e romano* (Messina: Hoepli, 1998); William Brashear, Francisca A.J. Hoogendijk, "Corpus Tabularum Lignearum Ceratarumque Aegyptiarum," *Enchoria*, 1990, 17:21–54; Elizabeth A. Meyer, "Writing Paraphernalia, Tablets, and Muses in Campanian Wall Painting," *American Journal of Archaeology*, 2009, 113:569–597; Boudalis, *Codex and Crafts* (cit. note 1), pp. 21–34.

6 Detail from *Mitis Sarni opes*, edited by Antonio De Simone, Salvatore C. Nappo (Napoli: Denaro Libri, 2000), p. 37.

their interplay with other media (fig. 1.1a). It shows the Flemish mystic Jan van Ruusbroec (1293–1381) writing down notes on a wax board in a forest, inspired by the Holy Spirit. These notes were later copied onto parchment upon his return to the monastery.[7] This situation was effectively summarized by Richard and Mary Rouse when they observed that "as a support for the written word, wax tablets had a longer uninterrupted association with literate Western civilization than either parchment or paper, and a more intimate relationship with literary creation."[8] From the Greek tragedian Aeschylus to John Locke's *tabula rasa* and beyond, passing through Plato, Aristotle, the Stoics and many others, wax boards have also been used as a powerful metaphor for the human mind. Indeed, Sigmund Freud was inspired by a special kind of this device when he wrote his essay on human memory entitled "Notiz über den Wunderblock."

A number of extant examples of wax boards demonstrate the key role they have also played in the fields of education and administration.[9] An illumination, dating from ca. 1312, provides an example for both contexts (fig. 1.1b). It depicts the name of John the Baptist being inscribed on a wax diptych on the occasion of his circumcision. Below this is a school scene, featuring a wax board with a handle (*tabula ansata*).[10] The National Gallery in London also has a painting, the *Pietà*, probably from the workshop of Rogier van der Weyden, which features a priest with a wax board tucked into his belt (fig. 1.2e).[11] This figure can be seen as a modern counterpart to the Mesopotamian incantation

7 Detail from the ms. Brussel KB 19.295–97, fol. 2v. (ca. 1380), public domain.

8 Richard H. Rouse, Mary A. Rouse, "The Vocabulary of Wax Tablets," in *Vocabulaire du livre et de l'écriture au moyen âge: Actes de la table ronde, Paris 24–26 septembre 1987*, edited by Olga Weijers (Turnhout: Brepols, 1989), pp. 220–230, p. 220.

9 On wax tablets in the Middle Ages and Modern Age, see Elisabeth Lalou, *Les Tablettes à écrire de l'Antiquité à l'Époque moderne. Actes du colloque international du CNRS, Paris, Institut de France, 10–11 octobre 1990*, Bibliologia 12 (Turnhout: Brepols, 1992); Reinhard Büll, *Das große Buch vom Wachs. Geschichte Kultur Technik* (München: Callway, 1977), pp. 785–894; Peter Gerlach, "Ein Lüneburger Wachstafelbuch aus dem 14. Jahrhundert," *Lüneburger Blätter*, 1965, 15/16:21–70; Kristina Krüger, "Schreibgriffel und Wachstafeln als Zeugnisse von Schriftlichkeit im Mittelalter," in *Text als Realie*, edited by Karl Brunner, Garhard Jaritz (Wien: Verlag Österreichische Akademie der Wissenschaften, 2003), pp. 229–61. Fig. 1.2d shows a waxed board-book from a monastery in or around Nuremberg, with notes and financial records (Staatsbibliothek Bamberg Msc.Var.15 [urn:nbn:de:bvb:22-dtl-0000017849]), dating to the 16th century CE.

10 Zürich, Schweizerisches Nationalmuseum, LM 26117, f. 179v — Gradual from St. Katharinenthal (Thurgau) (https://www.e-codices.ch/en/list/one/snm/LM026117).

11 © The National Gallery, London 2019, image no. NG6265. This kind of "pocket" wax board with annexed stylus was common in the Middle Ages, see e.g. Büll, *Wachs* (cit. note 9), p. 849 Fig. 629.

FIGURE 1.1 a) The mystic Jan van Ruusbroec (1293–1381) writing on a wax board. b) 14th-century illumination depicting a wax diptych and a *tabula ansata*. c) Modern tablet computer in context

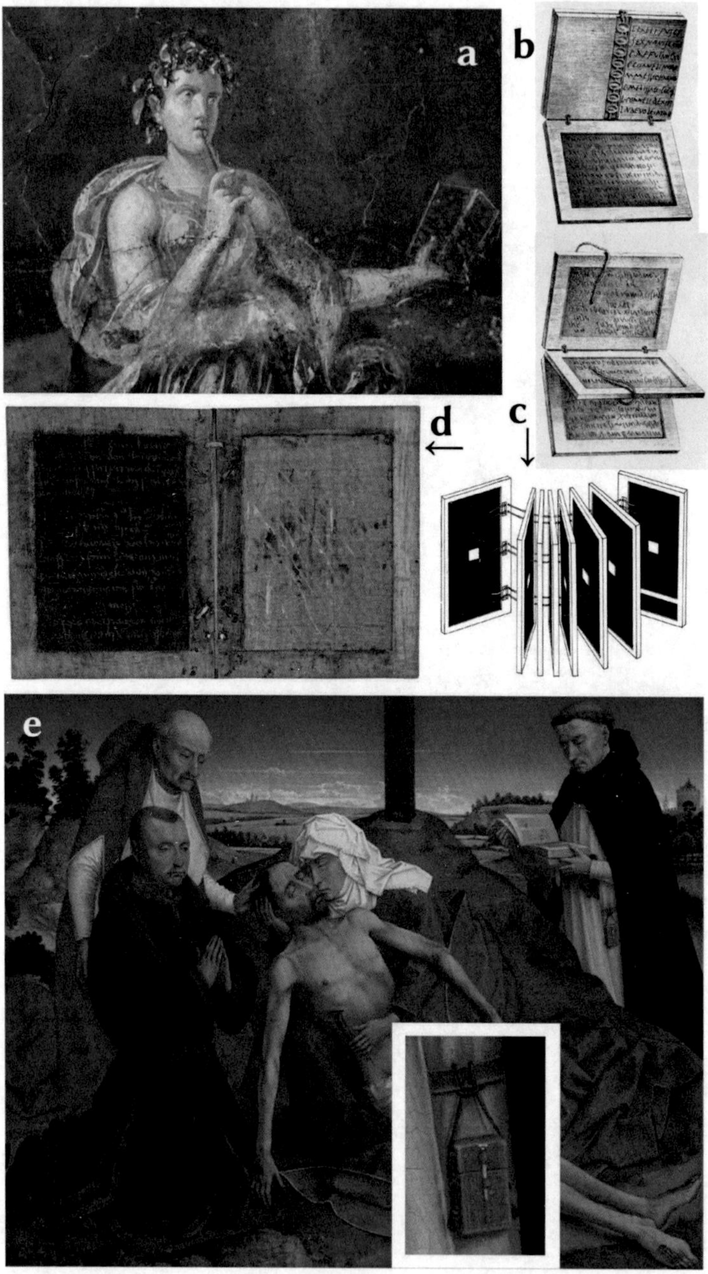

FIGURE 1.2 a) The Muse Calliope holding a wax diptych, from a Pompeii fresco. b–c) Reconstruction of a Roman wax board and of a multi-page board book. d) 16th century; wax board from a German monastery. e) A Pietà from the workshop of Rogier van der Weyden (15th cent.) depicting a priest with a wax board tucked in his belt

priest "carrying [his] writing board," mentioned in the "Poem of the Righteous Sufferer," a composition dating from the late second millennium BCE.[12]

3 Colour Matters

In both Classical times and the Middle Ages and, moreover, in modern Europe, the wax layer of most wax boards was black. The reason for this is that in those times beeswax was normally mixed with charcoal or soot in order to improve the plasticity of the paste. Since the letters were scratched into the wax layer with a pointed stylus, they were of the same colour as the black background, with the result that the legibility of such a board was much poorer than that of an inked piece of papyrus, parchment or, later, paper, especially in bad light conditions.[13] This is also clear from a passage by the Roman educator and rhetorician Quintilian: while recommending wax boards as the ideal medium for literary creation, he admits that text inked on parchment is more legible.[14] Other authors also complained about the difficulties of reading boards made from black wax paste (Martial, for example, describes them as *tristes cerae*), instead praising the quality of green and red wax pastes. This attitude is reflected in the poems of Baudri, abbot of Bourgueil from 1079 to 1106, whose peculiar compositions are obsessively concerned with the process of writing itself. In the poem *Ad tabulas*, he writes:

> Cera quidem vetus est, palearum fusca favilla,
> et turpat vestram cera vetus speciem.
> Idcirco minor est scribenti gratia vestra
> Cum velut offensum reiciat grafium.

---

12 *Ludlul bēl nēmeqi* III 41 (*mašmaššum-ma naši li'um*), see Wilfred G. Lambert, *Babylonian Wisdom Literature* (Oxford, 1960), p. 50.

13 Only occasionally did the stylus pierce through the wax of the writing board, so that the colour of the wood underneath showed through and heightened the contrast of the sign impression.

14 Quint. *Inst. Or.* 10.3.31: "scribi optime ceris, in quibus facillima est ratio delendi, nisi forte visus infirmior membranarum potius usum exiget, quae ut iuvant aciem, ita crebra relatione, quoad intinguntur calami, morantur manum et cogitationis impetum frangunt" ("It is best to write on wax owing to the facility which it offers for erasure, though weak sight may make it desirable to employ parchment by preference. The latter, however, although of assistance to the eye, delays the hand and interrupts the stream of thought owing to the frequency with which the pen has to be supplied with ink," quoted from *Quintilian. With An English Translation. Harold Edgeworth Butler* (Cambridge, MA & London: Harvard University Press & William Heinemann, 1922), pp. 108–109).

Ergo pro nigra viridantem praeparo ceram,
Quo placeat scribae gratia vestra magis ...[15]

And in another poem, *Ludendo de tabulis suis*:

Sit vobis oculos viridis color ad recreandos,
Sint indiruptae vincula corrigiae![16]

Indeed, the existence of both green and red boards is documented by textual, iconographical, and archaeological sources. For example, the wax of one of the oldest preserved Greek boards is of a reddish colour, as is that of a number of Roman tablets recovered at Murecine, near Puteoli;[17] green tablets are well attested in iconographical sources and there are a number of extant examples.[18] The technology of wax boards in the Greek and Roman world can only be investigated through analysis of archaeological finds and meagre hints in literary sources. There are "recipes" available, however, for Medieval and Early Modern European specimens. The most interesting manuscripts — masterfully collected and discussed by D. Oltrogge — date from the 15th century, among them the well-known *Liber illuministarum*.[19] By combining the evidence from these recipes with literary and iconographical sources and with the analysis published by Reinhard Büll in 1977, we get a varied picture of the

15 "As for the wax, it is old and black with grit, and this old wax disfigures your beauty. So you are less indulgent of the writer and resist his stylus as though you found it odious. Hence I am preparing green wax to replace the black, so as to make you more tolerant and friendly toward the scribe," translation after Roger Chartier, *Inscription and Erasure. Literature and Written Culture from the Eleventh to the Eighteenth Century* (Philadelphia, PA: University of Pennsylvania Press, 2007), p. 3.

16 "Let your colour be green, to regenerate the eye, let your straps be undisruptable!"

17 Martin West, "The Writing Tablets and Papyrus from Tomb II in Daphni," *Greek and Roman Musical Studies*, 2003, 1:73–92; Robert Marichal, "Les tablettes à écrire dans le monde romain," in Lalou, *tablettes à écrire* (cit. note 9), pp. 165–85, p. 171 fn. 34.

18 Lalou, *Tablettes à écrire* (cit. note 9), 233–88; Büll, *Wachs* (cit. note 9), 808–14. One of the most famous iconographical examples of a green wax board is found in the so-called *Gregorblatt* (Trier, Stadtbibliothek, Hs. 171/1626; 10th cent. CE). Stimulating observations on the role of green in the context of reading and writing practices are found in Leah Knight, *Reading Green in Early Modern England* (Farnham et al.: Routledge, 2014), pp. 28–30.

19 D. Oltrogge, "Wachsfärbung," in *Der ‚Liber illuministarum' aus Kloster Tegernsee: Edition, Übersetzung und Kommentar der kunsttechnologischen Rezepte*, edited by Anna Bartl, Christoph Krekel, Manfred Lautenschlager, Doris Oltrogge (Stuttgart et al.: Franz Steiner Verlag, 2005), pp. 658–62; *ead.*, *Kunsttechnologische Rezeptsammlung. Datenbank mittelalterlicher und frühneuzeitlicher kunsttechnologischer Rezepte in handschriftlicher Überlieferung* (TH Köln, Institut für Restaurierungs- und Konservierungswissenschaften, <http://db.re.fh-koeln.de:2200>).

ingredients used in the last two millennia to produce wax pastes for writing boards, including oils of various origin, resins and turpentine, dairy products, honey, ochre, charcoal, soot, verdigris, cinnabar, red lead (minium), azurite, and basic lead carbonate (white lead).[20] We can assume that black paste was generally cheaper than green or red paste, since charcoal was readily available, not so verdigris or minium, the production of which also required specialized knowledge in order to be successfully mixed with beeswax.

4 Back to Babylon

The history of wax boards predates Greek and Roman times, reaching back to the last quarter of the third millennium BCE. Indeed, the Greek word for "wax tablet," δέλτος (*déltos*), is a Phoenician loanword, ultimately derived from the Akkadian word *daltu*, meaning "door," also used in ancient Mesopotamia to indicate the leaf of a waxed board book.[21] The medium enjoyed growing popularity in the cuneiform cultures of Mesopotamia, Anatolia, and the Levant, with a peak in the Neo-Assyrian and Neo-Babylonian periods. In the Ancient Near East, wax boards appear to have been used primarily for writing cuneiform script and secondarily for linear scripts like Anatolian hieroglyphs and Aramaic. They were widely used for a range of different text genres, both documentary and literary.

The existence of wax boards alongside cuneiform clay tablets poses a question: what triggered the invention of wax boards in the Ancient Near East? Two aspects are problematic in this regard. Firstly, climatic factors meant that beeswax had mostly to be imported into Mesopotamia from neighboring regions. Secondly, cuneiform script is written by pressing the squared end of a stylus (mostly made of reed) into moist clay, i.e. it configures a writing technology that does not require ink and allows for the written text to be easily erased and corrected provided that the clay remains wet — precisely the features that have been defined above as the key assets of wax boards. There are two factors, however, that represent advantages of wax over clay, namely the possibility of adding text to an existing document over an extended period of time, and the ease of transport over long distances. It seems reasonable to

20 Oltrogge, *Wachsfärbung* (cit. note 19); ead. *Rezeptsammlung* (cit. note 19); Büll, *Wachs* (cit. note 9), pp. 796–820.

21 For evidence of wax boards in the Ancient Near East, see Cammarosano et al., *They Wrote on Wax* (cit. note 1).

assume that the invention of wax boards in Mesopotamia was triggered precisely by these factors.

The most ancient archaeological find pertaining to wax boards in an Ancient Near Eastern context comes from the Uluburun shipwreck and dates from the 14th century BCE, while the boards from the Palace of Sennacherib at Nimrud (near present-day Mosul, 8th century BCE) represent the only example of a board with a portion of the wax layer preserved and showing traces of cuneiform signs impressed into it (fig. 1.3). According to analyses performed in the 1950s in the laboratories of the British Museum, the wax layer of the Nimrud writing boards consists of beeswax compounded with ca. 25% orpiment (As_2S_3), a yellow mineral containing mainly arsenic sulfide, which is highly carcinogenic when swallowed or inhaled. This pigment can only be handled today in a chemical laboratory, a fact that prompted the start of the "WoW! Writing on Wax" project. Further evidence is found in cuneiform tablets from the Neo-Babylonian and Achaemenid periods (7th–6th century BCE), containing expenditure accounts related to the manufacture of wax boards. These texts not only mention wooden boards, but also regularly refer to beeswax and yellow ochre (Akkadian *kalû*) as the necessary raw materials, thus providing indirect evidence for the composition of the wax paste. One of these texts also mentions a certain quantity of sesame oil, yet its purpose is unclear (see below).

A bizarre circumstance further complicates the study of Ancient Near Eastern wax boards. Since wedges are produced simply by impressing a squared tip into a moist surface, the same stylus can be used to write cuneiform both on clay and wax. While not a single secure example of stylus for "cuneiform" wax boards is known, styli — both for clay and wax — are widely attested in iconographical sources, primarily as a symbol of the god Nabu in seal impressions and stelae, and within writing scenes in Neo-Assyrian wall panels. Curiously, a closer look at these sources highlights a fundamental difference in the appearance and handling of the styli connected with wax boards as compared to those used for clay tablets: while both kinds are of rectangular or trapezoidal shape, styli for wax boards display what resembles a longitudinal line or groove, whereas styli for clay tablets do not (fig. 1.4). In addition to the "groove," wax-styli sometimes also have one or more transversal bands. Particularly telling are the writing scenes found in Neo-Assyrian wall panels. Here, styli connected to writing cuneiform script on wax boards regularly show a groove on one side, and often also a horizontal band encircling them in the middle; the groove, however, is visible only when the scribes face left, otherwise only the horizontal band is shown. Interestingly, scribes with board-books are regularly shown holding the stylus upright and with index and middle fingers extended.

FIGURE 1.3 One of the sixteen ivory leaves hinged together in a folding board-book from Nimrud, 8th century BCE, and a detail of the wax layer preserving traces of cuneiform signs

FIGURE 1.4 Examples of "grooved styli" from two Babylonian *kudurru*-stones (BKR nos. 40 & 53), and from the Neo-Assyrian wall panel BM 124956

Since both the grooved and non-grooved styli are intended for cuneiform script, the difference must be related to the use of wax as opposed to clay as writing surface. What was the function of the alleged groove? In a recent article, Ursula Seidl suggested three possible options: the release of a pigment, of a substance preventing the stylus from sticking to the wax, or of a substance that softened the wax while impressing wedges.[22]

22 Ursula Seidl, "Assurbanipals Griffel," *Zeitschrift für Assyriologie und Vorderasiatische Archäologie*, 2007, 97:119–124, p. 124.

These hypotheses are strengthened by our reinterpretation of a passage of the tablet BM 28825, a late copy of a letter to king Assurbanipal, written on behalf of the scholars of Babylon in response to the king's request for scholarly cuneiform texts. In this letter, a reference to (date) syrup, ghee, and pressed (sesame) oil is found in fragmentary context following the mention of "seventy-two writing-boards of *sissoo*-wood"; most importantly, the ingredients are intended for "soaking" styli in a kettle:[23]

> These twelve scholars have, stored in their minds like goods piled in a magazine (i.e. they know off by heart), [the entire corpus of scribal learning that] they have read and collated, and the [..., they have toiled day and] night (writing it all down); they shall not shirk, from the property of the great lord Marduk, my lord, and all the houses in [... PN] my dear brother, who [...] seventy-two writing-boards of *sissoo*-wood from the [house (or temple) ...] he (or I) got out [...] syrup, ghee and pressed (oil) for the kettle of their styli, to soak (them into it), ([... *diš*]*pu*(làl) *ḫimētu*(ì.nun.na) *ù ḫal-ṣa ana ruqqi*(šen) *qan*(gi)-*ṭup-pí-sú-nu ana ṣe-pu-ú ú-še-ṣi*) and a one-litre vessel of the [...] he (or I) got out [for] their [...] and the chief scholar will exchange the tablet (credit-note?) for silver at Babylon.[24]

This fascinating passage suggests that styli used to write cuneiform on wax boards were occasionally soaked in a special mixture before writing. This raises a number of questions: in what proportions were these ingredients mixed together? What is meant by "soaking"? Are we to imagine a scribe repeatedly dipping the stylus in the oily mixture while writing? If so, what purpose did the substance serve? Is this reference related to the above-mentioned "groove," i.e. that the mixture referred to in the letter is precisely the substance that flow onto the writing surface while writing? Or, should we understand from this passage that the styli were to be soaked in preparation for writing?[25] If so, what purpose did such a procedure serve?

---

23 BM 28825 obv. 17. This passage has been kindly brought to our attention by M. Frazer. The translation offered by Grant Frame, in Grant Frame, A.R. George, "The Royal Libraries of Nineveh: New Evidence for King Ashurbanipal's Tablet Collecting," *Iraq*, 2010, 67:265–284, p. 275, requires modification: "he (or I) got out [oil,] syrup, ghee and pressed (oil) to soak the kettle of their styli"). Frame and George have responded positively to the interpretation proposed here (pers. comm.).

24 BM 28825 obv. 10–18, quoted after Frame and George, *Royal Libraries* (cit. note 23), pp. 274–75, with modifications.

25 The formulation in the text allows for both interpretations, that of "dipping" and of "soaking."

5 Wax Paste and Writing Technique

The evidence discussed above also opens up a number of technical questions. Was the wax paste of Ancient Near Eastern boards made solely of beeswax and yellow ochre or is it possible that it involved other ingredients, as was the case later? What is the specific effect of the various ingredients on the wax paste? Are the various pastes suitable for both cuneiform and linear scripts, or do script-specific constraints emerge? What was the function of the alleged "groove" of cuneiform styli for wax and what was the corresponding writing technique? These questions can only be investigated through an interdisciplinary approach in which the study of the sources and the chemical analysis are carried out in parallel. Consequently, we created many different pastes based on the available analyses, recipes from later periods, and our own hypotheses, and tested them systematically both for cuneiform and linear script under variable environmental conditions. In the following, a summary of the main results is presented, starting with an overview of the basic ingredients involved.[26]

5.1 *Beeswax*

Chemically, beeswax is a hydrophobic lipophilic substance — which means that, generally, it can be mixed with oily substances but not with water. With respect to the production of writing boards, the wax must be fluid enough to be poured into the recessed frame of the boards. Since it cannot be dissolved in water, either a non-polar solvent is needed or the solid wax has to be melted. There were no satisfactory results from our tests with historically available solvents. Beeswax melts between 62 and 65 °C since it is a mixture of many components,[27] most of which are long hydrocarbon chains that carry so-called functional groups. Different groups result in different chemical and physical properties in the molecules and therefore of the substances they form. Generally, beeswax can be considered as a mixture of mainly Monoesters, Hydrocarbons, Diesters and Hydroxy polyesters.[28] Esters account for 70% of the composition of beeswax, with myricylpalmitate being the most abundant.

The beeswax itself is the first thing that must be considered in any attempt to decipher a possible recipe for historical wax pastes. It is assumed that the

26 For a detailed discussion of these data see Cammarosano et al., *They Wrote on Wax* (cit. note 1).

27 Birgit S. Fröhlich, Wachse der Honigbiene Apis mellifera carnica Pollm.: chemische und physikalische Unterschiede und deren Bedeutung für die Bienen (PhD Diss., Würzburg University, 2000).

28 H. Randall Hepburn, Christian W.W. Pirk, Orawan Duangphakdee, *Honeybee Nests. Composition, Structure, Function* (Berlin et al.: Springer, 2014).

distribution of honeybees (*Apis mellifera*) in the Ancient Near East was largely the same as in modern times.[29] Our analysis, using IR-Spectroscopy, of beeswax from extant Egyptian honeybees (*Apis mellifera lamarckii*) and from the Würzburg colony showed no major detectable differences, hence our decision to use current German wax for our experiments. In our tests, pure beeswax proved to be rather unsuitable for writing cuneiform due to its stickiness and translucent texture (resulting in poor legibility), thus corroborating the written sources, which hint at the addition of further ingredients.

5.2 *Pigments*

The pigments we used for our experiments were ground to a fine powder using a mortar and pestle and then added to the molten beeswax. After being stirred to form a homogeneous mixture, the fluid was poured into the recessed frame of the experimental wooden boards. Once the paste had hardened, a standardized writing test was performed by a trained specialist.

A first optically evident effect of adding pigments is the modification of the writing surface's colour, with the orpiment causing a bright yellow and the ochre producing the typical brick red or brownish yellow colour. Thus, the addition of ochre produces pastes of a colour resembling that of clay tablets and orpiment imitates the appearance of gold.[30] It is no coincidence, therefore, that the luxury ivory boards from the royal Palace of Sennacherib at Nimrud contain orpiment, whereas the Neo-Babylonian accounts only mention ochre: the latter was no doubt the standard additive used in ordinary Mesopotamian boards. Charcoal and verdigris, commonly used in the Middle Ages and in modern Europe, were also tested, producing plain black or bright turquoise pastes, respectively. As discussed above, the colour of the paste is a key factor for the legibility of cuneiform and linear scripts, since in both cases it is strongly dependent on the contrast between the writing and the background in a given light. We can apply a general observation to both types of scripts, namely that the least legible background colours are white and black, whereas sienna, green and turquoise all perform well — an observation that supports the available written sources (see above, §3). In view of the fact that high quality ochre was readily available all over ancient Mesopotamia, it seems likely that yellow ochre was the standard additive to beeswax in Ancient Near

29 Cammarosano et al., *They Wrote on Wax* (cit. note 1), pp. 125–129

30 An interesting comparison is provided by the various techniques used in Medieval illuminated manuscripts to imitate the colour of gold, see Doris Oltrogge, "'Scriptio similis auri.' Gold und Goldähnlichkeit in der Handschriftenausstattung: Surrogat, Imitation, Materialillusion?," in *Codex und Material*, edited by Patrizia Carmassi, Gia Toussaint (Wiesbaden: Harrassowitz, 2018), pp. 159–178.

Eastern wax boards, whereas charcoal may have been used in Greece and Rome, simply because it was the most widely available pigment and subsequently canonized as a standard ingredient.

Apart from the colouring, the addition of pigments to the wax paste reduces the beeswax' stickiness and increases its malleability.[31] As far as ochre is concerned, two of the Neo-Babylonian cuneiform accounts mentioned above suggest a proportion of 6.6% and 10% of ochre in the wax paste, a quantity that is very close to the average amount of ochre present in the sample of wax boards analysed by Reinhard Büll.[32] Our tests showed that the addition of ochre (or arsenic sulfide),[33] up to an amount of ca. 40–50%, generally makes the paste softer than pure beeswax and reduces its stickiness, thus meeting the basic requirements for a writing board. The proportion of either of these ingredients may vary between ca. 5% and ca. 50%; pastes with more than 50% ochre become too granular and too hard, especially at low temperatures, so that the wedges can be impressed only with difficulty.

5.3 *Oil*

In the tests we conducted, adding sesame oil had the effect of making the wax more sticky and it produced a rather greasy consistency. Consequently, the pastes were not suited for writing cuneiform. However, oils are among the ingredients in modern recipes for wax board pastes, a fact which deserves further investigation.

5.4 *Resins*

Both ancient recipes and the analysis carried out by Reinhard Büll on 37 boards, dating from a period spanning from the Late Antiquity to the 18th century CE, attest to the occasional presence of resins (colophony and myrrh) and turpentine in the paste of wax boards.[34] This is puzzling, since our tests showed that the addition of resin hardens the wax paste, thus making it rather unsuitable for writing. We submit that the addition of resin may be related to the custom of mixing beeswax with both pigments and resins in sealing practices, and in some cases may also be connected to the quality of the wax that was used at that time.[35]

---

31 If the quantity of pigment is gradually increased, the paste gets harder again.

32 Büll, *Wachs* (cit. note 9), pp. 808–15.

33 Importantly, arsenic sulfide serves this purpose well, also when used alone, and necessitates no additional ingredients.

34 Büll, *Wachs* (cit. note 9), pp. 808–14; on the use of resins see also Oltrogge, *Wachsfärbung* (cit. note 19), pp. 585–90 and 595–96.

35 For more details see Cammarosano et al., *They Wrote on Wax* (cit. note 1), pp. 157–158.

5.5 *The Role of Physical Factors*

The temperature of the paste and the shape and physical properties of the stylus are important factors that influence writing cuneiform on wax. The effort required to impress wedges is directly proportional to the amount of pigment present in the wax paste, and inversely proportional to the temperature of the writing environment. This means that the more pigment the paste contains, the greater the pressure required to write; conversely, the higher the temperature, the easier it is to impress wedges.[36] Writing cuneiform on a paste containing 50% ochre at minus 10 °C requires a lot of pressure, whereas a paste with only 7% ochre feels very soft when writing at a room temperature of 35 °C.

5.6 *Writing Technique*

In order to investigate the hypothesis that Mesopotamian "wax styli" had a longitudinal groove, we tested various options, the most intriguing one being the idea of a "reservoir" for liquid substances that were intended to flow onto the writing surface continuously during the writing process (fig. 1.5). These substances either worked as a release agent, thus preventing the stylus from sticking to the wax surface, or they may have coloured the wedges, thus improving contrast and legibility. The latter option was not feasible, since the liquid could not properly flow along the appropriate corner of the stylus down into the wedges. That said, coloured wedges can be achieved when a larger amount of fluid containing a convenient quantity of date syrup (or other colouring substance) is used, and the surface is then wiped with a sponge (or similar), immediately after writing. While the part of the fluid that is on the surface is wiped off, the part which flowed in the wedges remains there and, after the oil has been absorbed, results in more or less red-coloured wedges, a sort of Mesopotamian counterpart to the *litterae rubricatae* of the Greco-Roman world (fig. 1.5). Whether such a technique was ever used in ancient Mesopotamia remains a matter of speculation.

The option of a groove to optimize the use of an oil-based release agent, of the kind referred to in the cuneiform letter BM 28825, proved to be more conducive. Our experiments involved periodically dipping the stylus in a viscose mixture composed of sesame oil, date syrup and ghee, while writing. These experiments showed that the use of such a release agent prevents the stylus from sticking to the surface and thus damaging the shape of the wedges, a problem arising especially under specific temperature conditions and when using bone or metal styli (fig. 1.5).

36 This empirical conclusion applies to mixtures of beeswax and ochre (up to 50%), and does not contradict the observation that such mixtures feel softer than *pure* beeswax when writing cuneiform on them.

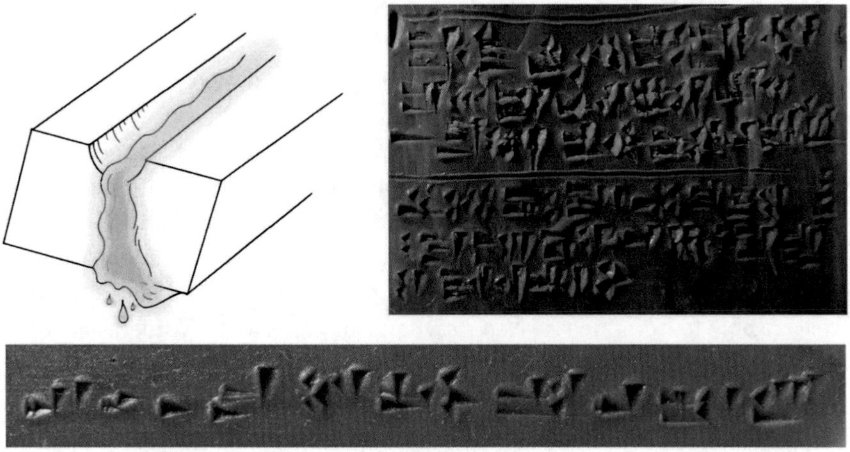

FIGURE 1.5 Schematic operation of an alleged "grooved stylus," coloured wedges achieved by means of a release agent with a high content of date syrup, and an example of the difference in the appearance of wedges obtained by means of a metal stylus without (top paragraph) and with (bottom paragraph) the use of a release agent made from sesame oil, date syrup and ghee

The use of an oil-based release agent has two drawbacks: the stylus has to be periodically dipped into it, and, unless one manages to permeate only the writing tip, it becomes slippery and greases the writer's finger. Both a proper groove cut along the main side of the stylus and a simple pattern carved on it may mitigate these drawbacks, as they function as a "reservoir" for the release agent (thanks to the effect of surface tension) and increase the grip on the stylus' shaft. Of course, the reservoir function is minimal in the case of engraved lines, maximal in the case of a proper groove. The pace at which the release agent flows onto the writing surface depends on its viscosity, hence on its components and on temperature, and is also influenced by the shape and dimensions of the groove (or lines) engraved on the stylus. These conditions seem to explain the use of ghee and syrup as components of the mixture attested in the cuneiform tablet BM 28825: while oil alone is able to attain the same desired effects, the addition of ghee and syrup gives the mixture the necessary viscosity.[37] When such a fluid is used, it forms a film that is progressively

37 A set of experiments related to the composition of the release agent is presented in Feline Maruhn, *Altorientalische Wachstafeln. Alte Medien neu erforscht* (Zulassungsarbeit zur Ersten Staatsprüfung für das Lehramt an Gymnasien, Julius-Maximilians-Universität Würzburg, 2017), pp. 98–108. Experimental "grooved styli" of metal and bone have been manufactured with the invaluable assistance of Miron Sevastre (Martin von Wagner-Museum, University of Würzburg), to whom we express our gratitude.

absorbed into the wax paste. Under normal temperature conditions, after a couple of days, the film is almost invisible to the eye.

The question remains of whether such a groove is really necessary for the use of a release agent. The answer is — probably not, although a groove of appropriate shape and dimensions optimizes the "reservoir" function. In the absence of a groove, the stylus must simply be dipped more frequently into the release agent. Be this as it may, how can we explain the transversal bands — sometimes one, sometimes more — that are occasionally visible on depictions of "grooved styli"? Important evidence in this regard emerged in May 2018, when we were able to examine two Neo-Assyrian ivory rods at the Vorderasiatisches Museum in Berlin.[38] The two rods, which were found in a grave together with a wax diptych, show, respectively, one and two engraved lines running around the "shaft" (fig. 1.7). We propose to interpret them as writing implements, and submit that the "groove" is an iconographic motif characterizing styli used to write on wax. The motif would consist of a line engraved in the middle of the stylus, which would symbolize a (wax) diptych in profile view (fig. 1.6). Such a motif would have the advantages of simplicity and functionality: an immediate allusion to wax boards, brilliantly obtained within the spatial constraints of an elongated parallelepiped, and also improving the fingers' grip on the stylus.

Thus, we are left with two different interpretations of the "groove," one that is mainly functional (a real groove as a reservoir for a release agent); the other mainly aesthetic/iconographic (a motif alluding to a diptych). The former remains speculative, whereas the latter is supported by iconographical variants showing multiple transversal bands, a fact that cannot be reconciled with anything other than a symbolic reference to a closed diptych in profile view. However, the two interpretations are by no means mutually exclusive. At present, we can neither establish whether and when the practical and iconographical functions may have coexisted, nor whether there was a phylogenetic relation between them, i.e. whether the invention of the motif was triggered by the practice of providing wax styli with a groove or vice versa.

To conclude this section, we present an educated guess about what a brand-new cuneiform wax board may have looked like (fig. 1.7).[39]

---

38 Inv. no. VA Ass 3545.3–4, see for details and discussion Cammarosano et al., *They Wrote on Wax* (cit. note 1), pp. 162–166. We are very grateful to Lutz Martin for granting us access to the collections of the *Vorderasiatisches Museum* and to Gert Jendritzki for his support during our stay in Berlin in May 2018 as well as for the stimulating discussions.

39 The reconstruction model of the diptych has been conceived and manufactured by Gert Jendritzki (*Staatliche Museen zu Berlin, Vorderasiatisches Museum*), see Gert Jendritzki, Matthias Streckfuß, Michele Cammarosano, "Technische und materialgerechte Rekonstruktion einer Elfenbein — Klapptafel aus Aššur," *Mitteilungen der Deutschen*

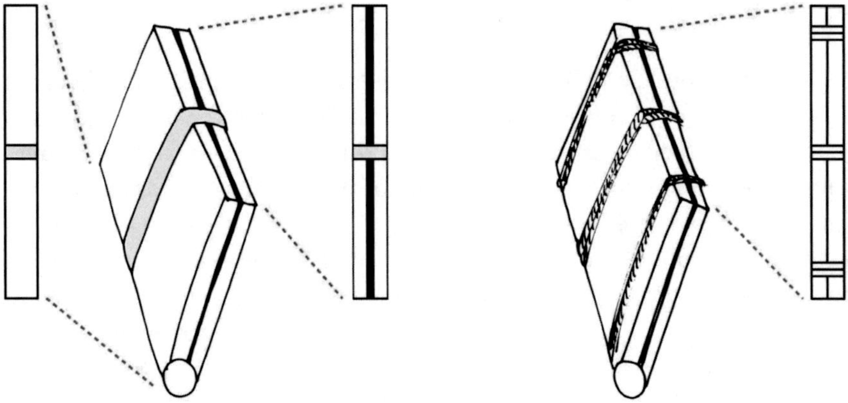

FIGURE 1.6 Schematic representation of the hypothetical gestation of the "grooved stylus" motif

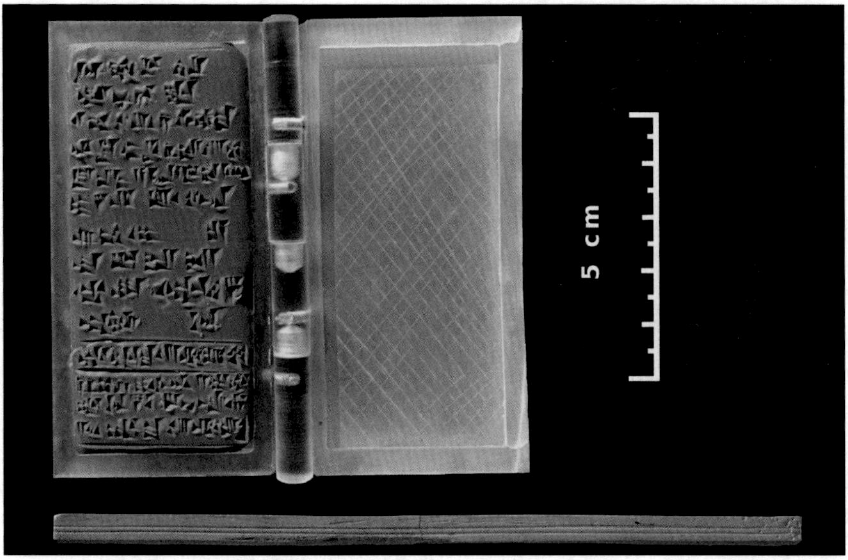

FIGURE 1.7 Top: The potency incantation LKA 95 rev. 6–11 written with a replica of the "grooved stylus" VA Ass 3545.4 on a Plexiglas reconstruction of the Neo-Assyrian diptych VA Ass 3541. Bottom: The "grooved stylus" VA Ass 3545.4

6 Wax Boards in Schools Today

As discussed above, wax boards have a long history in learning and schooling. Furthermore, one of the main aims of the "WoW! Writing on Wax" project is to communicate the research findings to the public, especially to schoolchildren, thus the idea of getting "writing on wax" back into schools is rather appealing. Since chemistry as well as history tend to be relatively unpopular subjects among pupils, an alternative approach is required.[40] One pedagogical approach to increase children's interest proposes non-formal, extracurricular settings for learning.[41] Museums, for example, offer the chance to explore objects that schools do not have access to. Such a direct encounter fosters a perception of authenticity that has proven to be helpful for learning.[42] Another approach is to encounter facts in a way that connects them to the everyday experiences or general interests of children. This adds an individual "sense-making" to the topics discussed, which, according to constructivist learning theories, supports their memorization.[43] Finally, studies show effective learning favours active doing over passive listening.[44] In terms of history, this may be the exploration

Orient-Gesellschaft 151 (2019), 201–218, also available at <www.osf.io/urpuf/wiki/home/>. The wax paste is made of beeswax with 25% ochre; the text reads: "Wind blow, orchard shake, Clouds gather, droplets fall! Let my potency be (steady as) running water, Let my penis be a (taut) harpstring, Let it not slip out of her!" (LKA 95 rev. 6–11, translation after B.R. Foster, *Before the Muses. An Anthology of Akkadian Literature* (Bethesda, 2005), p. 1000).

40 Jonathan Osborne, Shirley Simon, Sue Collins, "Attitudes towards Science: A review of the literature and its implications," *International Journal of Science Education*, 2003, 25:9, 1049–1079.

41 Katrin Engeln, *Schülerlabors: authentische, aktivierende Lernumgebungen als Möglichkeit, Interesse an Naturwissenschaften und Technik zu wecken* (Berlin: Logos Verlag, 2004); Fiona Affeldt, Sakari Tolppanen, Maija Aksela, Ingo Eilks, "The Potential of the Non-Formal Educational Sector for Supporting Chemistry Learning and Sustainability Education for all Students — a joint perspective from two cases in Finland and Germany," *Chemistry Education Research and Practice*, 2017, 18:13–25.

42 Ulrich Mayer, Peter Adamski, Klaus Bergmann (eds.), *Handbuch Methoden im Geschichtsunterricht* (Schwabach: Wochenschau Verlag, 2013); Jean Lave and Etienne Wenger, *Situated Learning: Legitimate Peripheral Participation* (Cambridge: Cambridge University Press, 2001); and Karen Malone, *Every Experience Matters: An evidence-based research report on the role of learning outside the classroom for children's whole development from birth to eighteen years* (Warwickshire: Face, 2008).

43 Ernst von Glasersfeld, *Was heißt „Lernen" aus konstruktivistischer Sicht?* in *Unterricht aus konstruktivistischer Sicht — Die Welten in den Köpfen der Kinder* (Neuwied: Luchterhand, 2002), and Kenneth G. Tobin, *The Practice of Constructivism in Science Education* (New York: AAAS Press, 1994).

44 Johanna Meixner, Klaus Müller, *Angewandter Konstruktivismus — ein Handbuch für die Bildungsarbeit in Schule und Beruf* (Aachen: Shaker-Verlag, 2004).

of historical objects, materials or ways of living. Conducting experiments is the favoured method for chemistry.

Considering these findings, two pedagogical programmes were developed. On the one hand, a set of experiments was designed that can replace or complement the existing, standard experiments in school. For example, esterification and saponification are usually explained using the example of oils or fats, but may be understood just as well using the example of beeswax esters being saponified when following a historical recipe for so-called Punic Wax. Professional characterization of waxes includes the determination of acid number, ester value and saponification number. In all three cases, the method was adapted to school experiments with beeswax. In order for this to make sense for the children, each experiment answers an authentic question in the context of writing on wax in history.[45] On the other hand, an extracurricular "lab day" at the Martin von Wagner-Museum in Würzburg was arranged and tested with a sixth-grade class.

7 A Cuneiform Lab

The main aim of the extracurricular programme was to inspire the sixth-graders and get them fascinated in writing on wax — and thus surreptitiously teach about life in the Ancient Near East — as well as to impart knowledge about chemical experimental competencies and the properties of common substances.[46] By approaching an authentic context from two perspectives — chemical and historical — a higher interconnectedness of knowledge and therefore a higher quality of learning can be achieved. Moreover, the lab provided a good opportunity to introduce schoolchildren to the riches of the Ancient Near East.[47]

---

45 Feline Maruhn, Katja Weirauch, Ekkehard Geidel, Michele Cammarosano, "Historische Wachstafeln — Alte Medien neu entdeckt," *Praxis der Naturwissenschaften — Chemie in der Schule*, 2017, 2:31–35.

46 A "call for applications" was launched in autumn 2016; class 6a of the *Riemenschneider Gymnasium Würzburg* was subsequently selected to participate in the Cuneiform Lab.

47 It must be stressed that the Ancient Near Eastern civilizations are not dealt with at all in today's German curriculum, so the children usually have no idea about that world. The situation is different in other countries, and it was different in the DDR (on which, see Hans Neumann, "Der Alte Orient in der Schule: Erfahrungen (und Perspektiven?) beim Verfassen von Geschichtslehrbüchern," *Altorientalische Forschungen*, 2016, 43:170–188. For other stimulating pedagogical approaches to the Ancient Near East, see Sabine Böhme, "Viele Wege führen nach Babylon. Das einjährige Museum & Schule-Projekt «Gartenzwerg trifft Nebukadnezar»," *Antike Welt*, 2016, 47/5:37–41; and Cécile Michel,

Pedagogically, the first step must always be the activation and verbalization of the learners' preconceptions. At the time the Cuneiform Lab was organized, the Martin von Wagner-Museum was hosting an exhibition on the Assyrian capital Nineveh, the site of which lies on the outskirts of present-day Mosul, in northern Iraq.[48] The exhibition focused on high-resolution, large format photographs of a number of scenes from the wall panels of king Sennacherib's "Palace without Rival," including those portraying scribes recording booty on wax boards. At the time of the exhibition, Mosul was sadly the scene of severe clashes between the Iraqi forces and ISIL, which were reported in the media and thereby provided an immediate access to the geographical area of interest, since most of the children had heard of the war in the Middle East. The photographs shown in the exhibition also depicted scenes from wars of that time, triggering a discussion about weapons, warfare and politics in the Ancient Near East — and today. Since there was a lot of information to master within a relatively short period of time, it was crucial to break down the contents and focus the children's interests. We therefore chose not to do a simple guided tour through the exhibition, but rather gave the children tasks to fulfil in small groups. For example, the groups were given several types of wooden swords and had to find out which one corresponded to those used by the Assyrian army, as portrayed in the wall panels (fig. 1.8). The children also discovered that the "correct" sword had a cuneiform inscription, a fact that served to introduce this kind of script.

While discussing the fascinating circumstances that led to Sennacherib's death,[49] the pupils asked how we know about this 3000-year-old crime. The image of two scribes led to the conclusion that there is written evidence — namely, cuneiform tablets that survive to this day. The next phase of the Cuneiform Lab took place in the rooms of the Department for Ancient Near Eastern Studies (adjacent to the museum), where the children had the opportunity to learn the cuneiform writing system. Coming into contact with

"L'écriture cunéiforme, première écriture dans l'histoire de l'humanité. À l'école des scribes," *La lettre de l'INSHS*, 2013, 25:8–12.

48 The exhibition *Ninive. Bilder aus Sanheribs „Palast ohnegleichen"* at the Martin von Wagner-Museum of the Würzburg University (21.10.2016-19.02.2017) was organized by the Department for Ancient Near Eastern Studies of the Würzburg University in cooperation with the *Centro Archeologico Ricerche e Scavi* of Turin. We are most grateful to Jochen Griesbach, director of the museum, for allowing us to organize the Cuneiform Lab in the framework of the exhibition.

49 Simo Parpola, "The Murderer of Sennacherib," in *Death in Mesopotamia, XXVIe Rencontre Assyriologique Internationale*, edited by B. Alster (Copenhagen: Akademisk Forlag, 1980), pp. 171–182.

FIGURE 1.8 Assyrian swords and ancient crimes: a moment during the Cuneiform Lab at the Martin von Wagner-Museum of the University of Würzburg

real cuneiform tablets creates an immediacy and authenticity not possible to achieve in school. We were able to bridge the time gap between the Ancient Near East and today by showing the pupils examples of Babylonian school children's attempts at writing cuneiform on clay tablets. With the help of a template with Babylonian cuneiform signs, the children were then challenged to write their own name in cuneiform on clay tablets, which could be taken home afterwards (fig. 1.9).

When writing on clay, the children experienced the difficulties of the clay drying out and becoming hard. The issue of clay tablets versus wax boards was discussed in relation to the wall panel that depicts two scribes visible in the exhibition. Beeswax was introduced as the only historically available material that allowed for repeated erasing or correcting of mistakes in writing over time. The problem of not knowing the recipe for wax tablets for cuneiform writing was also presented.

Moving into the lab rooms, the source of beeswax and the biology of the honeybee were explored. Finally, the main question of the "WoW! Writing on Wax" project was presented to the children: what is the ideal recipe for a wax paste for cuneiform writing? In order to come up with a potential recipe, the children first had to become familiar with the chemical and physical properties of historically plausible ingredients. This opened them up to chemical knowledge, rather than simply "learning" chemistry for no apparent reason, which

FIGURE 1.9 Learning cuneiform

schoolchildren often feel is senseless.[50] They conducted preliminary experiments to explore the solubility, melting point and thermal behaviour of possible ingredients for wax pastes. Using this knowledge while working in groups, they developed and documented their "ideal" recipe for a cuneiform-oriented wax paste (fig. 1.10).

Thus, by asking the same questions that the "real" scientists were working on and, like them, trying to find a solution, dealing with historical and chemical facts made sense. Additionally, on a meta-level the children learned a lot about a pedagogical construct that researchers relate to as the "Nature of Science" (NOS).[51]

A wide-ranging discussion has evolved in recent decades about what the aim of science education should be and what, consequently, should be taught.

50 Helena van Vorst, Sabine Fechner, Elke Sumfleth, "Kontextmerkmale und ihr Einfluss auf das Schülerinteresse im Fach Chemie," in *Inquiry-based Learning — Forschendes Lernen: Jahrestagung in Hannover 2012*, Gesellschaft für Didaktik der Chemie und Physik 33, edited by Sascha Bernholt (Kiel: IPL-Verlag, 2013), pp. 311–13.

51 See e.g. N.G. Lederman, "Students' and Teachers' Conceptions of the Nature of Science: A Review of the Research," *Journal of Research in Science Teaching*, 1992, 29/4:331–59; N.G. Lederman, "Nature of Science: Past, Present and Future," in *Handbook of Research on Science Education*, edited by S. Abell, N.G. Lederman, (Mawah: Lawrence Erlbaum Associates, 2007), pp. 831–79.

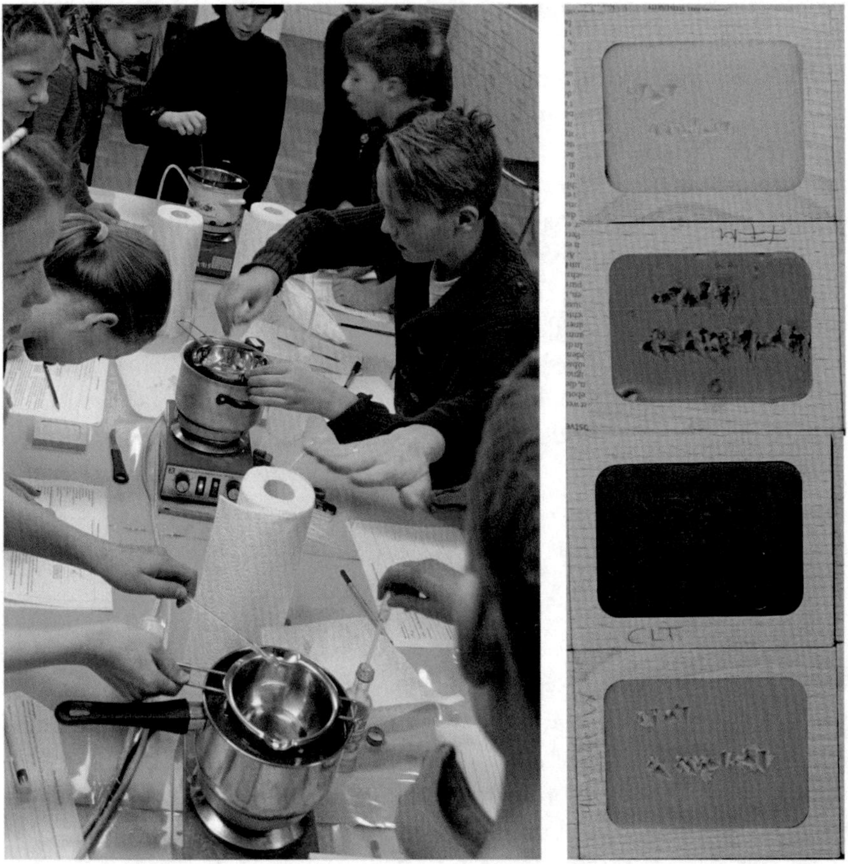

FIGURE 1.10 Mixing ingredients to obtain the ideal wax paste for cuneiform script, and a look at the results

As part of this, it was decided that every student should achieve "Scientific Literacy," enabling them to make rational decisions and become a responsible part of modern society.[52] An essential part of this literacy involves understanding what the nature of scientific knowledge is (NOSK), but also how this knowledge is achieved through inquiry (NOSI).[53] By acting themselves as researchers, the children experience important aspects and rules of the Nature of Science,

52 George E. DeBoer, "Scientific Literacy: Another Look at Its Historical and Contemporary Meanings and Its Relationship to Science Education Reform," *Journal of Research in Science Teaching*, 2000, 37/6:582–601; R.C. Laugksch, "Scientific Literacy: A Conceptual Overview," *Science Education*, 2002, 84/1:71–94.

53 Irene Neumann, *Beyond Physics Content Knowledge — Modeling Competence Regarding Nature of Science and Nature of Scientific Knowledge* (Berlin: Logos Verlag, 2011).

such as the need for preliminary exploration of the method, the difficulties of matching hypothesis and experiment, the importance of accuracy in documentation, but also — and most importantly — the fun of discovery. At the end of our lab day, every group handed in their wax board, and a specialist tested them by inscribing them in cuneiform, so that the best recipe could be identified — and rewarded with a chocolate equivalent.

No scientific inquiry has been done on the success of this pedagogical programme in terms of motivation and learning. However, since all the children — without exception — were working enthusiastically throughout the whole four hours, we dare to conclude that it is truly worth all the work and organizational effort.[54] The feedback from the three teachers involved confirmed this impression. Moreover, we proved that it is possible to successfully combine apparently distant topics, such as history and chemistry, within one pedagogical concept, resulting in powerful synergies not only for science, but for teaching as well.

8 Conclusions

Our search after ancient wax boards has become a fantastic journey at the intersection of archaeology, chemistry, cuneiform studies, and didactics. It is our hope that our study will shed some light on a rather forgotten writing technology that played an important role in the literacy of ancient times. Moreover, we hope that our trial will encourage others to embrace a truly interdisciplinary approach in the study of writing technologies, and to further explore ways of contaminating scientific projects with school teaching. Indeed, the main core of collaboration involved understanding each other's language and way of thinking (a process that continues to challenge us to this day), and finding ways to involve school children in the process of "making science." Their enthusiasm as well as the surprising scientific insights that arose in the course of the Cuneiform Lab have been among the greatest rewards for our efforts.

54 A report on the Cuneiform Lab is posted on the website of the *Riemenschneider Gymnasium* at <www.riemenschneider-gymnasium.de/2016/12/02/auf-den-spuren-der-keilschrift/> (last accessed 19 March 2020).

CHAPTER 2

Written in Blood? Decoding Some Red Inks of the Greek Magical Papyri

Miriam Blanco Cesteros

Abstract

In the procedures described within Greek magical papyri, it is common to find indications about the use of specific inks, usually characterized by the presence of peculiar substances. One such substance is blood, whose use is often interpreted in connection to the symbolic dimension of magic. A perusal of the relevant passages, however, reveals that some instances of "blood" have a different meaning. This paper analyses these formulas in the light of other sources in order to disclose the complicated network of *Decknamen* (code names) and to unveil the misunderstandings and erroneous inferences that have occurred in the course of their textual transmission.

Keywords

magic – inks – code names – papyri

Due to the large number of preserved magical texts and artefacts (amulets, lead tablets, ostraca, engraved gemstones, etc.), Greco-Roman Egypt offers one of the richest contexts for the study of the phenomenon of magic in Antiquity.* The significance of these documentary and archaeological witnesses lies in the

* This paper is part of the research project *AlchemEast — Alchemy in the Making: From Ancient Babylonia via Graeco-Roman Egypt into the Byzantine, Syriac, and Arabic Traditions*. The *AlchemEast* project has received funding from the European Research Council (ERC) under the European Union's HORIZON 2020 research and innovation program (G.A. 724914). I would like to thank Claudia Colini for providing me with some bibliography relevant to this paper, and also the other participants of the workshop *Traces of Ink* for their valuable suggestions that helped me improve my research on this topic. The last reviews of this contribution occurred after a change of academic affiliation to the Universidad Complutense Madrid.

fact that they provide direct evidence of the practice of magic in this period.[1] Among them, the so-called magical papyri stand out as the most important source of information. Modern scholars have classified under this name a very heterogeneous set of papyri,[2] not only because they spread over a large chronological frame — from the 2nd/1st century BC to the 5th/6th AC — but also because of the language in which they are written (Greek, Demotic and Coptic) and the varied character of their contents. An additional factor of heterogeneity is their nature. On the one hand, there are testimonies of "applied magic": writings and objects produced in the context of a magical ritual (e.g., magical gems, amulets and binding spells[3] written on papyri and lead tablets). On the other hand, there are texts intended to preserve and transmit the knowledge and practice of magic. Modern scholarship refers to these as magical handbooks because they contain collections of spells and magical practices described in varying degrees of detail.

The ritual instructions in many of these handbooks mention a large number of inks, sometimes even including specific indications for their preparation. These inks were used within the magical ritual to write the spell or a special formula, to draw the magical signs or the figures of the gods that should be depicted as part of the practice. When examining descriptions of their composition, a peculiar substance, frequently employed, stands out: blood. By

1 I will deploy the term "magic" and the adjective "magical" fully cognisant of the problems associated with them, merely as a conventional way to talk about a scattered body of artefacts, texts and practices, without prejudging its relationship with religion.

2 Leaving aside the individual editions of papyri with magical content, the first catalogue inventorying Greco-Egyptian magical papyri was by Karl Preisendanz (ed.), *Papyri Graecae Magicae Die griechischen Zauberpapyri*, 3 vols (Stuttgart: Teubner, 1928–1932), revised and republished by Albert Henrichs in 2 vols (Leipzig-Berlin: Teubner, 1973–1974). Hans Dieter Betz (ed.), *The Greek Magical Papyri in Translation including the Demotic Spells* (Chicago-London: University Chicago Press, 1986) offered an English translation that also included non-Greek parts — which were excluded from Preidendanz's edition — and new papyri. In the 1990s, Franco Maltomini and Robert W. Daniel (eds), *Supplementum Magicum* (Opladen: Westdeutscher Verlag, 1990–1991), edited two volumes of new materials. Since then, new texts are being regularly published by researchers. The magical texts collected by Preisendanz are usually cited using the abbreviation *PGM*, followed by the roman number referring to the papyrus position in that specific collection (e.g. *PGM* V). The texts edited in the *Supplementum* (abbreviated as *SM*) are referred to using roman numbers: e.g. *SM* 41. *PDM* stands for *Papyri Demoticae Magicae*. Regarding the use of editorial symbols in the quoted texts: round brackets insert interpretative specifications; square brackets indicate a lacuna; and angular brackets signify an unintentional omission by the scribe. Unless differently indicated, the English translations are mine.

3 This spell typology, which has several names in Greco-Egyptian magic (*katadesmos*, *katochos*, *agōgē*), served to restrain the will of divine and human beings in multiple contexts (erotic, judicial, agonistic contexts, etc.).

analysing the complex transmission of the magical texts and their codification using *Decknamen* (code names) this paper explores the possibility that not all the substances used in the magical inks called "blood" were, in fact, blood.

1 Understanding "Magical" Inks

Before engaging with specific cases of magical inks, it is necessary to understand what the context in which they were employed was.

Briefly, in the 19th-century magic was considered typical of primitive societies and lower-classes, a practice precursor to religion and opposed to it. These formulations, however, have been superseded. In Antiquity, magic was a widespread phenomenon, whose practice did not know differences of gender, religion or social status.[4] However, it is impossible to delineate a valid definition that may work both diachronically and synchronically, because its understanding varies significantly, depending on the point of view from which it is examined.[5] Usually, the ritual and medical practices of a culture become "magical" when perceived from a different perspective. In regard to Greco-Roman culture, for example, an examination of the terms "magic" and "magical" reveals that they often simply meant "foreign, different from the rituals of the public cult of the *polis*/state," without implying that they actually were "magical" according to modern standards. Certainly, the problem is not less complicated in Egypt. The Egyptian word usually translated as "magic" (*heka*) was actually a broad concept intertwined with medicine and, above

4 See, for example, Daniel Ogden, "Binding Spells: Curse Tablets and Voodoo Dolls in the Greek and Roman Worlds" in *Witchcraft and Magic in Europe: Ancient Greece and Rome*, edited by Bengt Ankarloo, Stuart Clark (Philadelphia: University of Pennsylvania Press, 1999), pp. 54–70.

5 This topic, which is a compulsory subject for any analysis of ancient magic or magical texts, has been the object of much scholarly discussion. Fundamental studies in this regard include: Jan N. Bremmer, "The birth of term magic," *Zeitschrift für Papyrologie und Epigraphik*, 1999, 126:1–12 and "Appendix: Magic and Religion," in *The Metamorphosis of Magic from Late Antiquity to the Early Modern Period*, edited by Jan N. Bremmer, Jan R. Veenstra (Leuven-Paris: Peeters Publishers, 2002), pp. 265–269; José Luis Calvo Martínez, "¿Magos griegos o persas? Los usos más antiguos del término *magos*," *MHNH*: revista internacional de investigación sobre magia y astrología antiguas, 2007, 7:301–314; Robert L. Fowler, "Greek Magic, Greek Religion," *Illinois Classical Studies*, 1995, 20:1–22; Fritz Graf, "Defining Magic — not Again?!," unpublished work whose draft is available at https://www.academia.edu/4054884/Graf_Magic (last accessed 10 Dec. 2018); Bernd-Christian Otto, "Towards Historicizing 'Magic' in Antiquity," *Numen*, 2013, 60:308–347; Henk S. Versnel, "Beyond cursing: the appeal to justice in judicial prayers," in *Magika hiera. Ancient Greek Magic and Religion*, edited by Christopher Faraone, Dirk Obbink (Oxford: Oxford University Press, 1991), pp. 60–106.

all, with religion. As a consequence, in the Egyptian socio-cultural context in which the Greek magical papyri were produced, these three fields were part of the same ritual reality and cannot be easily distinguished one from the other.[6] These are some of the reasons why many rituals recorded in the magical papyri are similar to religious ones from a formal perspective. Accordingly, although its definition remains debated by specialists, modern scholars propose that ancient "magic" is just another form of expressing the human relation with the divine, located in the same continuum as "religion,"[7] a field with which it was strongly interconnected.

Therefore, I will use the label "magical inks" to refer to those inks used in this particular context, although sometimes there is no difference between their composition and that of regular inks. In *PGM* VII 226, for instance, the ink employed is *melas graphikos* (μέλας γραφικός), that is to say, the usual black one. The majority of inks described in the Greek magical papyri, however, as produced in a ritual milieu with the purpose of functioning in a ritual context, are influenced by the performative nature of the magical procedure. This affects their composition, preparation and use.

Nowhere is this more evident than in ink recipes. Two kinds of components are fundamental to the composition of any ink: the pigment and the binder, which makes the ink adhere to the writing surface and guarantees the durability of the writing. Magical inks, however, can also contain peculiar components that do not make sense from a practical perspective. By contrast, their use can only be explained in terms of their symbolic value within a ritual framework: both the origin of these substances and the quantity in which they are employed respond to ritual requirements. This is the case, for example, with the "*7 wings of the Hermaic ibis*," as we can read in the recipe below, which is used to write a formula to strengthen memory:

> This is the preparation of the ink: myrrh *troglitis*, 4 drachmas, 3 Karian figs, 7 pits of Nikolaus dates, 7 dried pine cones, 7 piths of the single-stemmed wormwood, 7 wings of the Hermaic ibis (*ibeōs Hermaikēs ptera*, ἴβεως Ἑρμαϊκῆς πτερά), spring water. After burning (*kaysas*, καύσας) them, prepare and write.
>
> *PGM* I 243–246

6 See Geraldine Pinch, *Magic in Ancient Egypt*, 1st ed. 1994 (London: British Museum Press, 2010); Robert Kriech Ritner, *The Mechanics of Ancient Egyptian Magical Practice*, 1st ed. 1993 (Chicago–Illinois: Oriental Institute of the University of Chicago Press, 2008), pp. 3–28, 235–249.

7 Christopher Faraone, "The agonistic context of early Greek binding spells," in *Magika Hiera* (cit. note 5), pp. 17–20.

Without going into specifics, many cultures consider the number seven to be the bearer of a special symbolism because of different reasons (mythological causes, philosophical reinterpretations, religious beliefs, etc.).[8] In Greco-Egyptian magic, its symbolism is bound to so many cultural inputs and systems of thought that it is difficult to know its meaning with any certainty, other than the general view that it was considered as a special and powerful number. As for the ingredient itself, the ibis was identified with the god Thoth,[9] who, in turn, was associated with the Greek god Hermes.[10] From this syncretism the god Hermes-Thoth emerges. Hence the ibis is given the adjective *Hermaikē* (Ἑρμαϊκή) in the recipe and, in the magical practice, this ink is called the "ink of Hermes" (l. 246). Regarding the connection between Thoth and Hermes and the ink, these "civilizer" gods were creators of different knowledge and learning. As a consequence, they were also considered as the gods presiding over memory, as can be evinced from some texts on Hermes.[11] The inclusion in the recipe of a component associated with Hermes-Thoth established a sympathetic link between the divine power and the purpose of the formula — to improve memory — that guaranteed its efficacy.

From a merely technical point of view, this ink, prepared from the ashes of burned components, can be classified as a carbon ink, one of the oldest and most frequently used inks in Antiquity.[12] However, in the strongly ritualized context of magical practice, it is difficult to distinguish the act of burning ink's ingredients from the burning of an offering. In fact, the verb employed to describe the burning of ink's ingredients in some recipes[13] is *epithyō* (ἐπιθύω), a verb from the ritual sphere that means "to burn as an offering, to make a burnt

8 See, e.g., Milena Bogdanović, "The number and its symbolism in ancient Greece," *Journal of Arts and Humanities*, 2013, II 6: 116–121 and Rodney Ast, Julia Lougovaya, "The Art of Isopsephism in the Greco-Roman world," in *Ägytische Magie und ihre Umwelt*, edited by Andrea Jördens (Wiesbaden: Harrassowitz, 2015), pp. 82–98.

9 See, e.g., Donald B. Redford, *The Oxford Encyclopedia of Ancient Egypt*, 3 vols (Oxford: Oxford University Press, 2001), *s.v.* "Thoth."

10 Garth Fowden, *The Egyptian Hermes* (Princeton: Princeton University Press, 1986), pp. 23–31.

11 See, e.g. *h.Merc.* 428–429; *h.Orph.* 28.12 and Call. *Iambi*, Fr.221 Pfeiffer.

12 See the summary, with specific bibliography, in Thomas Christiansen, "Manufacture of Black Ink in the Ancient Mediterranean," *Bulletin of the American Society of Papyrologists*, 2017, 54:167–195, pp. 171–175.

13 E.g. *PGM* IV 2205 and 2226. The preparation of the latter is disseminated among the instructions for a brief erotic practice. The recipe for this ink is thus difficult to distinguish at first sight. It is a kind of myrrh ink — the technical term in *PGM* is *zmyrnomelan* (ζμυρνόμελαν) — based on myrrh and wormwood (common ingredients in magical inks), burned with sumac (a substance with high-tannic content employed in Antiquity to produce pigments) and roses. Since roses were considered a creation by Aphrodite, this last ingredient was probably a sympathetic one.

offering." This suggests that this action could function simultaneously on both levels, i.e. the making of the ink and the magical ritual.

The special, ritual application of these inks also explains why an essential — and functional — component of ordinary inks, i.e. the binder, is absent from many of them.[14] A large number of magical inks were used to write texts that were meant to disappear (by the action of time, water, fire, etc.) rather than endure for a long time. This was also a simple way to ensure a practice was impossible to annul. In other cases, the ink acted as a vehicle for the magical properties of the formula, transferring them to the practitioner who had to lick the text or drink the water that had been used to wash the text.[15] Since, in these cases, it was not intended for the pigments in these inks to adhere to the writing surface in a durable way, the binding agents were superfluous.

However, leaving aside the performative context, in terms of their technical characteristics, the procedure for the production of magical inks was no different from that used to make non-ritual inks: the ingredients were reduced into powder by burning or grinding and then mixed with a moistening agent in order to dissolve it or to create a suspension. Similarly, the majority of the ingredients used in magical inks (e.g. wormwood, pine cones, figs, myrrh, etc.) can be found in other ancient recipes for the preparation of inks, as Thomas Christiansen has recently demonstrated.[16] According to Christiansen's research, all the inks of *PGM* are classifiable within the three categories of inks known in Antiquity:[17] the aforementioned carbon inks, iron gall inks[18] and mixed inks.[19] After these preliminary remarks, it can be stated that an ink is defined as "magical" by the context in which it was produced or used, rather than by any technical reasons related to its composition or preparation.

14 In fact, only one recipe (*PGM* XII 97–99, cited below) explicitly mentions gum arabic (*kymmi*, κύμμι) as ingredient. However, its use can be considered implicit when a recipe includes common black ink, e.g. *PGM* VIII 69–73.

15 See, e.g., *PGM* I 232.

16 Christiansen, *Manufacture* (cit. note 12).

17 I follow the classification given in Christiansen, *Manufacture* (cit. note 12).

18 An ink made by adding some iron sulphate to a solution of tannic acid, a compound extracted from oak galls. The tannic acid reacts with iron salts producing a brown-black ink, see Christiansen, *Manufacture* (cit. note 12), p. 170 with specific bibliography.

19 Carbon inks that add metallic components to the mixture; in other cases, they are made from a suspension of some metallic powder in a binding substance, see Christiansen, *Manufacture* (cit. note 12), p. 170. Such inks are also called compound inks.

2 Written in Blood

The expression "written in blood" becomes literal in Greco-Egyptian magic, which had a particular preference for red inks, the preparation of which sometimes included the use of blood as a peculiar and characterizing ingredient.

Whatever the culture, blood is a substance that has had and, indeed, continues to have an undeniable symbolic and spiritual value: no other substance has ever represented the power of life and death like blood. It is perhaps for this reason that ancient cultures actually had an ambivalent attitude towards it, considering blood to be simultaneously sacred and a bearer of danger and impurity.[20] As a ritual sphere, Greco-Egyptian magic also shared this dual approach to the vital humor: on the one hand, blood was employed as an offering for the gods as well as for consecrating objects;[21] on the other, the spilling of different kinds of blood was denounced as a crime in magical practices.[22] Its main use, however, was as an ingredient for ink (see Table 2.1).

We should note that, in Antiquity, the use of blood as (or in) ink was neither restricted to magic, nor was it necessarily related to ritual purposes. Firstly, blood was the simplest and most immediate way to obtain a red substance to write with. On the other hand, different sources inform us about the use of blood as (or mixed with) ink in non-magical texts.[23] When used in iron gall inks, for instance, its use is entirely practical: the high iron content of the blood

20 Regarding ancient Greece, for example, animal blood was poured onto the altars during blood sacrifices; purification by blood was considered among the most effective rituals of this kind, see Robert Parker, *Miasma: Pollution and Purification in Early Greek Religion*, (Oxford: Oxford University Press, 1983), p. 230. But while the blood cleaned and released worshippers from their sins, its spilling implied a murder, which tainted whoever committed it (*ibid.*, pp. 104–143). In fact, any bloodshed was dangerous, likewise life events linked to blood, such as childbirth and menstruation (*ibid.*, pp. 270–273).

21 E.g. on blood used for consecrating the space and objects for the magical practice, see *PGM* II 160; IV 2888; VII 868; XXXVI 233; on blood used as offering, see, e.g. *PGM* II 177 and IV 2883.

22 The context of these denunciations is always a ritual *diabolē* (slander spell) in which the magician denounces the actions and words of someone to the gods in order to ensure a divine punishment for this person. Such actions, however, were not committed by the denounced persons (for this reason it is slander). By contrast, some of the practices reveal that the denunciations were actually done by ritual practitioners, who later attributed them to the victim of the spell. These are, e.g. a goat's sacrifice in *PGM* IV 2575 (=IV 2643), done by the magician at l. 2685; the spilling of donkey's blood in *PDM* xiv, done by the magician at ll. 679–681.

23 See, e.g. the manuscript studied by Marina Bicchieri and Flavia Pinzari, "Discoveries and oddities in library materials," *Microchemical Journal*, 2016, 124:271–273.

TABLE 2.1 Blood as an ingredient of ink and other uses in Greco-Egyptian magical texts

Ink	– The written text serves as protection against *daimons* (spirits): *PGM* IV 79 – to invoke a *daimon* as *paredros* (assistant): *PGM* IV 1999, 2094, 2096, 2098, 2210; XIa 2 – to invoke a *daimon* for a *katadesmos* (binding spell): *PGM* VII 301, 652; XXXVI 71 (Typhon), 362; LXI 61; LXII 104 – to invoke a god with oracular purposes: *PGM* VII 222 and VIII 69 (Bes); XII 145 (Hermes); XIII 315 (unknown) – unknown aim: *PGM* XXXVI 264	60%
Other uses	II 160, 177; IV 41, 2883, 2888, 3257; V 2666; VII 868; XIb 4; XXXVI 233; *SM* 100a, l. 4	36,6%
Dubious	*PGM* III 410[a]	3,3%

a The text is very lacunose, but as it is immediately followed by the indication "write with a pen," this blood was probably used as ink or ink component.

reacts with tannic acid, darkening the ink.[24] This same chemical reaction is also behind the utilization of blood for invisible inks:[25] texts written using diluted blood could be revealed as a result of the chemical interaction with other substances like the aforementioned tannic acid.

Another interesting point that emerges from the table above is the strong connection of inks made with blood to harmful magic and, in general, the demonic world. This specialization could be due to its colour. Indeed, in ancient Egyptian culture red was a powerful and ambivalent colour associated with many things that were considered both positive and dangerous (like, e.g., fire or blood itself). Its negative overtones seem to originate from its association with the desert and its sovereign, the god Seth. Consequently, this colour assumed the negative ideas that the desert and Seth personified, including the notions of "evil,"[26] chaos, hostility, disorder, violence and pain.[27] As a result,

24 As proven by laboratory replication, cf. Bicchieri, Pinzari, *Discoveries* (cit. in the previous note), p. 272.

25 Such use is mentioned regarding early Medieval Jewish handbooks, by Martin Levey, "Some Black Inks in Early Mediaeval Jewish Literature," *Chymia* 1964, 9:27–31, p. 28.

26 Ritner, *Mechanics* (cit. note 6), p. 147, n. 662.

27 See Herman T. Velde, *Seth, God of confusion* (Leiden: Brill, 1967), pp. 81ss.; Redford, *The Oxford Encyclopedia* (cit. note 9), p. 269, *s.v.* Seth; Pinch, *Magic* (cit. note 6), pp. 191–194, *s.v.* Seth. The basic word for red, *dšr*, was closely connected with that of anger and fury, both ideas also personified by Seth, see Geraldine Pinch "Red things: the symbolism of colour

red was associated with demonic figures and hostile gods;[28] it was also used for writing the names of such divinities and enemies, while red rubrics were avoided for other gods.[29] Therefore, considering the Egyptian background of the magical papyri, the use of red inks in aggressive magic and in the invocation of Underworld entities, such as spirits of the dead, does not seem arbitrary. Despite this, as stressed by Geraldine Pinch, "red things were dangerous, but they might, if handled in the right way, provide the most powerful protection available,"[30] so red could be also employed as an effective defense against any threat.

In my opinion, however, the symbolic and ritual dimension of the magical papyri has conditioned the interpretation of blood in magical inks. Modern studies about the inks described in Greco-Egyptian magical handbooks limit any consideration of blood to it being a ritual component. Consequently, the presence of blood in these texts has not received the attention it deserves: its presence is always explained — quite superficially — by the magician's wish to ensure the efficaciousness of the spell.[31] Without excluding the actual use of blood in Greco-Egyptian magical rituals, the passages discussed below will evince that some mentions of animal bloods in ink recipes from Greco-Egyptian magical texts could actually be concealing the existence of completely different substances.

3 Snake Blood

The first case study concerns a substance called *haima drakonteion* (αἷμα δρακόντειον). It was used as an ingredient in an ink meant for writing a series of *nomina magica* in a procedure intended to conjure the spirit of a dead person as *paredros* (assistant). The recipe is as follows:

> The ink: snake's blood and goldsmith's soot (αἷμα δρακόντειον καὶ αἰθάλη χρυσοχοϊκή)
>
> *PGM* IV 1999

in magic," in *Colour and Painting in Ancient Egypt*, edited by W. Vivian Davies (London: British Museum Press, 2001), p. 184.

28 Ritner, *Mechanics* (cit. note 6), p. 147–148, especially note n. 662 and 663.

29 *Ibid.*, p. 147, note n. 663 with specialized bibliography.

30 Pinch, *Red things* (cit. note 27), p. 184.

31 See, e.g. Christiansen, *Manufacture* (cit. note 12), p. 180.

The adjective *drakonteion/drakontion* (δρακόντειον/δρακόντιον) stems from *drakōn* (δράκων, a kind of snake), meaning "of/from the *drakōn* snake" or, in a more general way, "of/from the snake," so *haima drakonteion* is, literally, "snake blood."[32] Several ancient sources, however, inform us that, despite its name, this was a vegetal substance. One of these sources is the *Cyranides* (1st/2nd century AD),[33] a six-book treatise on the magical and medicinal properties of animals, plants and stones. In a passage entitled "the correct identification of the Snake-Plant, the Woodpecker Bird, the Dragon-Fish and the Arboreal Stone (δρακόντιος βοτάνη, δενδροκολάπτης πτηνόν, δράκων ἰχθύς, δενδρίτης λίθος εὔγνωστος)" this treatise states:

> From the squashed seeds of this plant [i.e. the Snake-Plant], [a substance] is obtained that is called "snake blood" because it is red.
>
> *Cyran.* 1.4.8 Kaimakis

The physician Aetius of Amida (5th/6th century AD) concurs with the *Cyranides* and, without specifying the part of the plant from which it originates, repeats that the *snake blood* is produced "by the Snake-Plant in the Indian lands" (Aët. XV, 13.141 Zervos). In fact, even today, "dragon's blood" is the common name of a bright red resin exuded by different kinds of plants such as the *Dracaena cinnabari*.

On the other hand, it should be noted that both Aetius and the *Cyranides* explain what *snake blood* was; this suggests that the name of this substance probably led to confusion and that untrained people required clarification about the real nature of this substance. In this regard, Pliny the Elder (23–79 AD) is perhaps one of the most interesting witnesses to the ancient misinterpretation of what the *snake blood* was:

> The Greeks give also to minium (i.e. the cinnabar, in lat. *minium*) the name of *cinnabaris*; a mistake is made because of this Indian name, for they give this name to the blood of a snake (*saniem draconis*) crushed

32 On the term *drakōn*, its meaning in Antiquity and for a general overview of its semantic evolution, see Daniel Ogden, *Drakon: Dragon Myth and Serpent Cult in the Greek and Roman Worlds*, (Oxford: OUP 2013), pp. 2–5. On the substance called *haima drakonteion*, with special attention to its botanical identification and the analysis of the sources, see Jean Trinquier, "Cinnabaris et 'sang-dragon': le 'cinabre' des anciens entre minéral, végétal et animal," *Revue archéologique*, 2013, 56:305–346.

33 On the complex tradition of this treatise and the discussion about its dating, see David Bain, "Περιγίνεσθαι as a medical term and a conjecture in the *Cyranides*," in *Ethics and Rhetoric: Classical Essays for Donald Russell* edited by Doreen Innes, Harry Hine *et al.* (Oxford: Clarendon Press, 1995), pp. 281–286.

> by the weight of dying elephants, which die mixing their blood with the blood of those animals — as we said — and there is no other colour that represents the blood in paintings more adequately.
>
> PLIN., *nat.* XXXIII, 116[34]

Pliny's explanation (most thoroughly referred in *Nat.*, VIII 12) surely is a popular belief originated from the literal interpretation of the expression *haima drakonteion*, which Pliny translates as *saniem draconis*. On the other hand, Pliny's account is interesting because it mentions a terminological misunderstanding between two red substances: namely, *snake blood* and cinnabar (HgS). The latter was called *kinnabari* (κιννάβαρι) as well as minium (*minion*/μίνιον), but certain literary sources, like the *Periplus Maris Erythraei* §30 (ca. 40/70 d.C.), evince that the *snake blood* in Antiquity was also called "(Indian) cinnabar." This led to the confusion between the two substances, as the Greek physician Dioscorides (c. 40–c. 90 AD) also discusses in Dsc. V 94.[35] In this passage, in fact, the author distinguishes between *minion* (with which he refers to cinnabar) and the *cinnabari*, which is, according to him, a very rare Libyan substance that he describes as follows:

> It has an intense colour (βαθύχρουν), whence some people believed that it was snake's blood (ὅθεν ἐνόμισάν τινες αὐτὸ αἷμα εἶναι δρακόντιον). The *cinnabari* has the same properties as the hematite; it is particularly suitable for ophthalmic remedies ...
>
> DSC. V 94

Although the text is ambiguous with respect to its formulation, two reasons lead me to consider that Dioscorides is here reflecting the same popular misinterpretation as Pliny and, therefore, in this passage, *haima drakonteion* should be taken to be the actual blood of a snake.[36] Firstly, Dioscorides talks about an erroneous popular belief (*enomisan tines*, he says); secondly, and most importantly, Dioscorides appears not to know of the vegetal substance called

34 Translation by Matteo Martelli, "Properties and classification of mercury between natural philosophy, medicine and alchemy," *AION* (Philol.), 2014, 36:17–48, p. 40. This translation follows the Latin edition by Zehnacker, Hubert, *Pline l'ancien: Histoire naturelle, livre XXXIII* (Paris: Les Belles Lettres, 1983).

35 About the confused and misleading terminology using in Antiquity for both the *snake blood* and cinnabar, as well as on a detailed analysis of the Pliny's and Dioscorides' texts here quoted, see Martelli, *Properties* (cit. note 34), pp. 39–44 and Trinquier, *Cinnabaris* (cit. note 32), 305–320.

36 For a radically different interpretation, see Pietro-Andrea Mattioli, *I Discorsi nei sei libri di Pedacio Dioscoride Anazarbeo della materia medicinale* (Venezia, 1568), pp. 1415–1416.

snake blood, since this is the only mention in his work. In any case, Dioscorides' quotation confirms that, in Antiquity, there was a certain confusion regarding the denomination and identification of several red pigments, a phenomenon undoubtedly underpinned by its chromatic character.[37] In fact, two recipes transmitted in alchemical treatises support that *snake blood* was used as a pigment, as Pliny says.

The first is a procedure for producing a red gemstone,[38] preserved in the Stockholm Papyrus, a Greco-Egyptian alchemical handbook (4th century ca.) written by a scribe who also produced an important magical handbook (*PGM* XIII). The second, although transmitted by a later alchemical treatise — the *Mappae clavicula* (henceforth *MC*) — provides the most meaningful evidence for the interpretation of the recipe from *PGM* IV 1999 (the magical ink with which we began this analysis). The Latin alchemical recipe book, compiled circa the 8th century, probably originated from a lost Greek original that collected a tradition of recipes very close to that of the Stockholm Papyrus; that is to say, another Greco-Egyptian treatise from roughly the same period.[39] This is the recipe:

> Add Indian *snake blood* to the gold (*sanguine draconis indici inunge aurum*) and put them in a copper container. Place charcoal around it and leave it there until they [i.e. the two aforementioned substances] melt and it [i.e. the mixture] is so liquefied that you can write with it.
>
> *MC* XLV

I want to stress the similarity between the ingredients employed in this ink — *snake blood* and gold — and those of the magical ink — *snake blood* and "soot of goldsmith" (*aithalē chrysochoikē*, αἰθάλη χρυσοχοϊκή). The latter, however, is difficult to interpret. *Aithalē* (αἰθάλη) commonly means "soot, thick smoke."[40] This term was used, in natural philosophy and medicine, to describe

37 Note that echoes of such identification remain in the botanical name of the tree *Dracaena cinnabari*.

38 P.Holm. §62 (l. 422).

39 The numerous Graecisms of the oldest versions prove the dependence of the *Mappae clavicula* on a Greek source. Although this Greek model has some recipes in common with P.Holm. and its brother-treatise, the P.Leid. X, it has recently been argued that the MC's source was not a compendium derived from P.Holm. and P.Leid. X, but rather a contemporaneous treatise belonging to the same alchemical tradition, see Gaia Caprotti, "Mappae clavicula: prescrizioni della prima alchimia storica nei precedent di lingua greca," in *Mappae clavicula. Alle origini dell'alchimia in Occidente* edited by Sandro Baroni *et al.* (Milano: il Prato, 2013), p. 219.

40 See *LSJ* (the *Liddell and Scott Greek–English Lexicon*), *s.v.* αἰθάλη.

substances that condense in a solid form (i.e., as a soot) during (or as a result of) the treatment of some ores.[41] Yet, no derivative is obtained when gold is heated. Given the parallels with *MC*'s recipe, however, we cannot rule out that the "soot of goldsmith" from the magical ink was simply gold or some kind of golden blend described in a technical or, more probably, hermetic way.[42] If so, both the magical and the alchemical recipe would simply employ a red resin with gold.

Other possibilities emerge, however, from another passage from the magical papyri: *PGM* XII 410. This text once again features the three ideas just discussed — i.e. the misleading nature of the name of this substance; the misinterpretations that this caused; and the confusion of *snake blood* with other red minerals for chromatic reasons:

> snake's blood: hematite stone (αἷμα ὄφ[ε]ως· αἱματίτης λίθος)

This quotation belongs to a list — to which I shall return in a moment — of alleged *Decknamen* (code names) used by magicians to conceal the real name of the substances used in their rituals. A quick glance at any dictionary reveals that this *haima opheōs* (αἷμα ὄφεως) is our *haima drakonteion* (αἷμα δρακόντειον): *ophis* (ὄφις), which also means "snake," is an interchangeable synonym for *drakōn* (δράκων)[43] that has replaced *drakonteion* here. In this context, *drakonteion* is a botanical name that refers to the 'snake-plant.' Its replacement with *opheōs* ("of snake") transforms the phrase *haima drakonteion* (i.e. a red substance that comes from the 'snake-plant') into real and literal snake blood. Since this is a list of equivalences (see below), it implies also that the magician considered *snake blood* a name for hematite. This stone, whose name stems from the Greek word *haima* (αἷμα, "blood") due to its coloration,[44] was also mentioned by Dioscorides in the passage quoted above because of its similarity with *snake blood*.[45] The magical text thus proves that *snake blood* and hematite could be erroneously identified as a single substance.

41 E.g., mercury obtained from cinnabar, or cadmia from copper, see Martelli, *Properties* (cit. note 34), pp. 27–32.

42 Similarly, e.g. in the alchemical tradition, "goldsmith's glue" (*kolla chrysochoikē*, κόλλα χρυσοχοϊκή) is the name of a type of golden alloy (e.g. P.Leid. x §32 and 44).

43 See *LSJ*, *s.v.* δράκων.

44 I agree with Caley-Richards that, when ancient authors talk about the hematite like a bright red gemstone, they did not mean the metallic gray mineral known until today as hematite, see Earle R. Caley, John F.C. Richards, *Theophrastus, De Lapidibus/Theophrastus, On Stones* (Ohio: Ohio State University Press, 1956, repr. 2016) pp. 138–139. See also George Rapp, *Archaeomineralogy* (Berlin-Heidelberg: Springer Verlag, 2009), pp. 113–114.

45 See above. See also note 35 for bibliography.

This distortion of both the common technical name and the real nature of the substance illustrates the kind of changes and alterations that technical terms could experience in the context of fluid traditions, like the one that characterized ancient recipes.

4 Blood of Donkey/Typhon

The second case of inquiry, while different, illustrates similar phenomena. It is the "blood of donkey," a substance employed for writing (as ink or mixed with other ingredients to make an ink) and mentioned in the following passages from *PGM*:

> [a.] Take the blood of a black ox or of a goat or of Typhon — but preferably from a goat — and write.
>
> *PGM* VII 652

> [b.] Formula written on a little piece of papyrus using Typhon's blood.
>
> *PGM* LXI 61

The donkey, when mentioned in connection to Egyptian culture, evokes an immediate association with the god Seth.[46] It was, in fact, one of the possible theriomorphic manifestations of this god. From the Hellenistic period, however, when Seth syncretized with the Greek monster Typhon,[47] the latter also assumed the attributes and forms of Seth, to the point that in Greek magical papyri the nouns "donkey" and "Typhon" are interchangeable.[48] Consequently, in Greco-Egyptian magic, we can consider the "blood of Typhon" (*haima Typhōnos*, αἷμα Τυφῶνος) and the "blood of donkey" (*haima onios*, αἷμα ὀνίος) to be the same substance. Accordingly, three more inks can be added to those mentioned above:

> [c.] Taking a piece of pure (καθαρόν, new?) papyrus write the following names with donkey blood (αἵματι ὀνίῳ).
>
> *PGM* XXXVI 71

46 Redford, *The Oxford Encyclopedia* (cit. note 9), p. 270.

47 Pinch, *Magic* (cit. note 6), pp. 193–194.

48 The equation donkey-Typhon is especially clear in *PGM* XI 2, which prescribes the use of a skull "of Typhon," clearly a donkey's skull. See also *PDM* XIV 675–694 (commented below and cited also above, in note 22).

[d.] Taking a seashell, write (on it) the sacred names with the blood of a black donkey (διὰ αἵματος ὀνίου μελάνου).

PGM VII 300

[e.] The hide is written using blood from the heart of a slaughtered donkey (αἵματι ὀνείῳ ἀπὸ καρδίας ἐσφαγμένου), which is mixed with the soot of coppersmith (αἰθάλη χαλκέως).

PGM IV 2093[49]

Unlike *snake blood*, which can quite safely be identified with a common botanical name, the substance called "blood of Typhon" or "blood of donkey" — at least at first sight — was the blood of this particular animal. This is evinced from, for example, the practice described in *PDM* xiv 675–694, in which the magician announces to the gods that certain people must be punished because they have "spilled the blood of Typhon in their houses" (l. 692). This assertion is equated with the magician's ritual of anointing his hands with donkey blood (l. 680).[50] The "blood of Typhon" could also refer to real blood in a passage like the example (a), in which the blood of other animals is given as an alternative. However, one of the magical inks listed above leaves some room for doubt about the general assumption that the "blood of Typhon" is always the blood of a donkey. It is the example (b) in which the "blood of Typhon" is glossed:[51]

Λόγος γραφό]μενος εἰς [χα]ρτίον[]/[αἵμα]τι Τυφῶνος, ὅ ἐστ[ι ...]

Formula written] on a little piece of papyrus using blood of Typhon, which is [...]

In the context of Greco-Egyptian magic, where the identification between Typhon-Seth and the donkey was so strong that their names are interchangeable, a gloss specifying what "blood of Typhon" was would have made no sense unless there were other substances also bearing this name. Unfortunately, the second element of the equation has been lost.

In the flexible system of ancient Greco-Egyptian magic, it is difficult to know what these other substances would have been, but there are several

49 I will return to this recipe.

50 On this form of curse, see n. 22, above.

51 Other passages in which a gloss is introduced use the formula *ho estin* x (ὅ ἐστιν ..., "which is ..."), see, e.g. *PGM* III 425 and XIII 128. It must not be mixed up with the use of this same phrase to introduce ritual identifications.

candidates. One could be, for example, wine. Plutarch (*de Iside* 6) reports that the Egyptians did not drink wine or use it as an offering because grape juice was considered "the blood of those who once were enemies of the gods" (ὡς αἷμα τῶν πολεμησάντων ποτὲ τοῖς θεοῖς). As Seth was the gods' enemy par excellence, some scholars identify the "blood of Seth" mentioned in some Egyptian rituals with wine.[52] The same is true of water from the Nile.[53] However, to my knowledge, neither wine, nor Nile water were used to write Greco-Egyptian magical texts.

There was a second substance associated with Typhon-Seth, much less peculiar than the donkey's blood, but also of red colour and commonly employed as a pigment: minium ("red lead" or lead tetroxide;[54] called *miltos* — μίλτος — in Greek).[55] In fact, *miltos* is mentioned among the ingredients of some magical inks:

> [f.] The ink: minium (μίλτου), burned myrrh and sap of fresh wormwood.
> *PGM* IV 2153

> [g.] [Write] with myrrh spiced with minium (μίλτῳ) on leaves of hellebore.
> *PGM* XIXb 2

> [h.] Take minium [and the blood][56] of a white dove (λαβὼν μίλτον ⟨καὶ αἷμα⟩ περιστερᾶς λευκῆς) or of a crow, as well as the 'milk' of the sycamore,

52 See Dierk Wortmann, "Das Blut des Seth," *Zeitschrift für Papyrologie und Epigraphik*, 1968, 2: 227–230.

53 *Ibid.*

54 According to the aforementioned texts of Dioscorides and Pliny (see §3, above), in Antiquity this mineral was commonly confused with other red pigments, such as cinnabar and iron oxides like hematite (red ochre). See also Rapp, *Archaeomineralogy* (cit. note 44), pp. 210 and 214.

55 See Rapp, *Archaeomineralogy* (cit. note 44), p. 214.

56 Karl Preisendanz integrated ⟨καὶ αἷμα⟩ before περιστερᾶς on the basis of *PGM* VIII 69–73, which transmits an alternative version of this magical procedure: "This is the ink with which you draw: blood of a crow, blood of a white dove (αἷμα περιστερᾶς λευκῆς), lumps of incense, (etc.)." One could argue that this second version does not include minium, so μίλτον περιστερᾶς λευκῆς (in *PGM* VII) could be replacing αἷμα περιστερᾶς λευκῆς (*miltos*, thus, would mean in this case "blood"). Some translators, in fact, reject Preisendanz's integration and interpret *miltos* in this way, see note 59 below. As it will be shown later, the minium could ritually replace the blood of a donkey in some Typhonian contexts. However, in order to create such a metaphor (using term *miltos* for the word *haima*, "blood," even when not related to Typhonian animals), the ritual substitution should have been much more frequent. Yet, the true is that, for now, it is attested only in one case.

> sap of single-stemmed wormwood, cinnabar, and rainwater; blend, put aside and write with it.
>
> *PGM* VII 222–226

The use of minium for writing, though, was not restricted to magic: according to the results of modern chemical analysis, it is, indeed, one of the commonest substances used in the red inks of Greco-Roman Egypt.[57] In Greco-Egyptian magic, however, it acquired a particular Sethian meaning,[58] surely due to its red colour, as proved, for instance, by passages in which the minium is called "minium of Typhon" (μίλτος Τυφῶνος/μιλτάριον Τυφῶνος):[59]

> Writing made with Typhonian ink (Τυφωνίου μέλανος γραφή): red anemone, juice from an artichoke, seed of the Egyptian acacia, Typhonian minium (μίλ[τ]ου Τυφῶνος), asbestos, quicklime, wormwood with a single stem, gum arabic, rainwater.
>
> *PGM* XII 97–99

> For spells that restrain (κατόχων): write on a seashell (εἰς τὸ ὄστρακον ἀπὸ θαλάσσης) with the ink mentioned below (i.e. the myrrh ink described in the next preparation), adding Typhonian minium (μιλτάριον Τυφῶνος).
>
> *PGM* IV 2211–2215

Several recipes for red inks transmitted in medieval Arabic treatises include red anemone (ἀνεμώνη) as the main ingredient, which does not prevent this flower from being "regarded as containing the concentrate essence of redness."[60] As

57 See Christiansen et al., "Chemical characterization of black and red inks inscribed on ancient Egyptian papyri: The Tebtunis temple library," in *Journal of Archaeological Science: Reports*, 2017, 14:208–219 and pp. 216–217.

58 The negative, Sethian connotations of the colour red — undoubtedly inherited from the Egyptian context — could explain, e.g. that certain practices stated the use of lamps "without minium" (*amiltōton*, ἀμίλτωτον) in order to avoid the dangerous influence of this colour on the ritual. For a complete list of texts, see Betz, *Magical Papyri* (cit. note 2), p. 336.

59 *Miltarion* (μιλτάριον) is a diminutive of *miltos*, which here can be interpreted as a mere lexical variant. I would like to stress that both expressions (μίλτος/μιλτάριον Τυφῶνος) have been interpreted in the quoted texts by some translators as "blood of Typhon," see Betz, *Magical Papyri* (cit. note 2), p. 77; and Luis Muñoz Delgado, *Léxico de magia y religión en los papiros mágicos griegos,* (Madrid: CSIC, 2001), *s.v.* μίλτος. These translations are based on *PGM* VII 222, which is interpreted according the papyrus' reading without Preisendanz's integration (see note 56, above).

60 Pinch, *Red things* (cit. note 27), p. 183. The mentioned recipes are transmitted in the *Zīnat al-kataba* (d. 925) by Abū Bakr Muḥammad ibn Zakariyyā' al-Rāzī; *Kitāb al-azhār fī 'amal*

far as minium is concerned, it developed a dual role as pigment and sympathetic ingredient. Therefore, both ingredients link the ink to the Sethian sphere by means of the colour. This is the reason, for instance, why the former ink is called "Typhonian."

Although minium is never called "blood of Typhon" in *PGM*, its red colour makes this equation likely. In fact, *PGM* VII 300–302 — a parallel version of *PGM* IV 2211–2215 —[61] demonstrates that, from a ritual perspective, minium and donkey's blood were commutable substances. Accordingly, this practice prescribes donkey blood instead of minium:

> A restraining procedure (ἀγώγιμον) that acts within an hour: take a seashell (ὄστρακον θαλάσσιον) and write the holy names on it with the blood of a black donkey (αἵματος ὀνίου μελάνου).

In sum, although the "blood of Typhon" is usually only another way of saying "blood of donkey," passages such as *PGM* LXI 61 prove that there were probably other substances that could also be called "blood of Typhon." In fact, the blood of donkey (that it is to say, the "blood of Typhon") could be replaced by minium in the ritual, as the last example demonstrates. This makes this pigment a good candidate for resolving the damage gloss of *PGM* LXI 61. Whether the minium could also have been called "blood of donkey" will be discussed in a moment.

---

al-aḥbār (13th century) by Muḥammad Ibn Maymūn al-Marrākušī al-Ḥimyarī; *Tuḥaf al-ḫawāṣṣ fī ṭuraf al-ḫawāṣṣ* (d. 1307) by Abū Bakr Muḥammad al-Qalalūsī. See an Italian translation of the recipes in Sara Fani, Le arti del libro secondo le fonti arabe originali. I ricettari arabi per la fabbricazione di inchiostri (sec. IX–XIII): loro importanza per una corretta valutazione e conservazione del patrimonio manoscritto, PhD Diss. (Naples: University of Naples, 2013), p. 43 R IV = p. 114 MḤ IV 2.d = p. 142 Q I.18; p. 106 MḤ III 4. On the scientific identification of the anemone see also *ibid.*, p. 218. See also Sara Fani's chapter in this volume "The Literary Dimension and Life of Arabic Treatises on Ink Making."

61 Recipes are texts of fluid tradition and their comparison involves scholars analysing three aspects to determine a link between the different versions of a text: aim, procedure and core ingredients. See Lucia Raggetti, *ʿĪsā ibn ʿAlī's Book on the Useful Properties of Animal Parts. Edition, translation and study of a fluid tradition* (Berlin: De Gruyter, 2018). In this perspective, the parallel between the two procedures is clear: the typology of both spells is the same — binding spells — and both must be written on a seashell using an ink that contains a component associated with Typhon.

5 The *Decknamen*

If one had to choose a text that best exemplifies the caution needed by scholars when approaching magical substances, it would be the aforementioned list of *PGM* XII (ll. 407–435). This list, which itemizes more than 40 names supposedly used to encode mineral and vegetal substances, is introduced as follows (ll. 400–407):

> *Interpretations from the translations of the temples employed by the Sacred Scribes* (ἑρμηνεύματα ἐκ τῶν ἱερῶν μεθηρμηνευμένα, οἷς ἐχρῶντο οἱ ἱερογραμματεῖς).
>
> Because of the curiosity of the masses, they (i.e. the Sacred Scribes) inscribed (ἐπέγραψαν) the names of the herbs and other things they employed on the statues of the gods,[62] so that, when interested in these matters, they (i.e. the masses) could not inquire about (περιεργάζωνται) any of them as a result of the error (i.e. the misunderstanding resulting from ignorance about the correct interpretation — *hermēneuma*, ἑρμήνευμα). We, however, have collected the solutions (i.e. to the names of the herbs and other things used by the Sacred Scribes) from many copies of these texts (ἀντιγράφων), all of which are secret.

The list that follows enumerates thirteen substances named "blood of *x*":

409	αἷμα ὄφ[ε]ως· αἱματίτης λίθος	Snake blood: hematite stone
411	αἷμα χοιρ[ο]γρύλλου· ἀληθῶς χοιρογρύλλου	Blood of hyrax; truly of hyrax
414	αἷμα κυνοκεφάλου· αἷμα καλαβώτου	Blood of baboon: blood of lizard
416	αἷμα Ἡφαίστου· ἀρτεμισία	Blood of Hephaistos: wormwood
419–22	αἷμα Ἄρεως· ἀνδράχνη	Blood of Ares: purslane
	αἷμα ὀφθαλμοῦ· ἀκακαλλίδα	Blood of an eye: tamarisk
	αἷμα ἀπ' ὤμου· ἄκανθις	Blood from a shoulder: bear's breach
	ἀπ' ὀσφύος· ἀνθέμιον	From the loins: chamomile

62 The phrase τὰς βοτάνας καὶ τὰ ἄλ[λ]α, οἷς ἐχρῶντο, εἰς θεῶν εἴδωλα ἐπέγραψαν, however, could be interpreted also as "they ascribed (i.e. assigned) the names of the herbs and other things they employed to the figures of the gods," see Miriam Blanco Cesteros "Los *Hermeneumata* de PGM/PDM XII (=GEMF 15): la *Dreckapotheke* mágica a examen" in *Cuadernos de Filología Clásica. Estudios Griegos e Indoeuropeos* 2020, 30:169–170.

426	Ἑστίας αἷμα· ἀνθέμιον	Blood of Hestia: chamomile
428	αἷμα χηναλώπεκος· γάλα συκαμίνης	Blood of a goose: a mulberry tree's 'milk'
431	αἷμα Κρόνου· κεδρίας	Kronos' blood: of cedar (resin of cedar?)
434	ἀπὸ Τιτᾶνος· θρίδαξ ἀγρία	Of Titan: wild lettuce
435	αἷμα ἀπὸ κεφαλῆς· θέρμος	Blood from a head: lupine

According to the introduction of this text, then, the substances itemized in this list were employed by the Egyptian Sacred Scribes (*hierogrammateis*, ἱερογραμματεῖς) — the Greek translation of an Egyptian sacerdotal category — who encoded their names to conceal their true nature to the uninitiated. The fact that these supposedly code names were inscribed in the temples by the priests leads us to the Egyptian Temple of Dendera. In this holy compound, there is a chamber called the "Goldsmiths' workshop," a ritual and artisanal space where the effigies of the gods were first produced and later consecrated. Philippe Derchain identified several inscriptions on the chamber walls, probably copied from treatises dealing with the administrative and practical operations of this workshop.[63] One of these texts[64] is a list of equivalences between the names of several materials and the actual nature of these substances. This passage was likely extrapolated from a treatise about the production of statues. Thus, we have archaeological evidence of the actual existence of inscriptions like those mentioned by the texts of *PGM* XII. According to Derchain,[65] however, Dendera's equivalences seem to be glosses of archaic or obsolete terminology, rather than code names. This is, precisely, the conclusion of Jacco Dieleman regarding the list of *PGM* XII: the names of the left column seems to be botanical names employed by Egyptian priests (above all physicians, as some medical texts prove) for which the right column provides a Greek identification. Therefore, this list might not be a list of code names, but a glossary of botanical terminology in two different languages.[66]

Either way, *PGM* XII list, motivated by secrecy or otherwise, illustrates three types of names depending on the provenance of the substance: names associated with an animal (like "snake's blood" and "donkey's blood"); with a god (such as the already analysed "Typhon's blood"); or with a body part (regarding

63 Philippe Derchain, "L'Atelier des Orfèvres à Dendara et les origines de l'Alchimie," in *Chronique d'Égypte*, 1990, vol. 65, n. 129: 219–242.

64 For the complete text, see Derchain, *L'Atelier* (cit. in the previous note), p. 235, text No. II.b.

65 Derchain, *L'Atelier* (cit. note 63), p. 223.

66 Jacco Dieleman, *Priests, Tongues, and Rites: The London-Leiden Magical Manuscripts and Translation in Egyptian Ritual, 100–300 CE.* (Leiden: Brill, 2005), pp. 185–203. This hypothesis has been recently ratified in Blanco Cesteros, *Los Hermeneumata* (cit. note 62).

inks, the "blood of the hand" is used, e.g. in *PGM* IV 79). This list is also interesting because, apart from supporting the idea that actual blood was used to write in Greco-Egyptian magic, some of these alleged *Decknamen* can be identified with substances mentioned in other magical texts. It is the case of the "blood of baboon," indicated as a writing medium in *PGM* XIII 315. Although, according to the list, it would be blood of lizard, Lynn R. LiDonnici has convincingly demonstrated, through comparison with other *PGM* passages in which baboon's *materia magica* is required, that the term *kynokephalos* ("baboon") in several cases replaced the term *cynocephalion* (κυνοκεφάλιον) — the common name of a plant.[67] Her conclusions, thus, require us to be cautious regarding the equivalence given by the list, which could be inaccurate, like in the case of the *snake's blood*. In any case, it is an additional piece of evidence that several mineral or vegetable substances — not necessarily for chromatic reasons — were referred to as blood in magical texts.[68]

6 Re-thinking the "Blood" of the Red Inks in Magical Texts

We can now analyse one last set of inks. They are mentioned in the same recipe, which indicates that the magical practitioner should write on three different materials, each time using a particular ink:

> Write on it [i.e. the hide] with blood of the heart of a slaughtered donkey (αἵματι ὀνείῳ ἀπὸ καρδίας ἐσφαγμένου), which is mixed with coppersmith's soot (αἰθάλῃ χαλκέως). The leaf of flax has to be inscribed with falcon blood (αἵματι ἱερακείῳ), which is mixed with goldsmith's soot (αἰθάλῃ χρυσοχόου); the *hieratic* papyrus has to be inscribed with eel blood (αἵματι ἐνχέλεως), which is mixed with acacia (ἀκακία).
>
> *PGM* IV 2093–2097

There is, thus, a first ink [INK A] made with blood of donkey and "coppersmith's soot;" INK B is made with the blood of falcon and "goldsmith's soot" and

67 Lynn R. LiDonnici, "Beans, Fleawort, and the Blood of a Hamadryas Baboon: Recipe Ingredients in Greco-Roman Magical Materials," in *Magic and Ritual in the Ancient World*, edited by Paul Mirecky and Marvin Meyer (Leiden: Brill, 2002), pp. 359–377, especially pp. 371–372. Note that this replacement is very similar to what has happened with *drakontion/ophis* in the case of the *snake's blood*. In fact, in her article, LiDonnaci evinces phenomena very similar to those pointed out in the present study.

68 See, e.g. the chamomile (called "blood of Hestia"). As for the name of the wild lettuce ("blood of Titan"), see Blanco Cesteros, *Los Hermeneumata* (cit. note 62), pp. 162–165.

INK C, with blood of eel and acacia. As is clear, the ingredients of these three inks immediately recall the first ink analysed in this paper, i.e. the ink made with *snake blood* and "goldsmith's soot" (αἷμα δρακόντειον καὶ αἰθάλη χρυσοχοϊκή) (henceforth INK D). The resemblance does not seem accidental, given that INK D appears in the same papyrus, *PGM* IV.

With respect to the substances employed in these inks, only the *snake blood* and the acacia can be identified with some certainty. The latter probably refers to the acacia' sap or *kymmi* (κύμμι); this substance, nowadays known as gum arabic, was commonly employed as a binder in the production of inks.[69] It is noteworthy, however, that the recipe refers to a well-known substance not by its customary name, but using an elusive term. This recalls the probable identification of the "goldsmith's soot" — INKS B and D — as simple gold (see above, §3). Can we assume, by analogy, that the "coppersmith's soot" of INK A is simply copper? It is difficult to say with certainty, but there is some indication that we should not interpret the name of these substances literally. Firstly, the "blood of falcon" stands out because it is the only mention in the Greek and Demotic magical papyri of its use as ingredient. There were, in fact, ritual reasons not to use the blood of this animal, which was identified with the solar god Horus and was, therefore, one of Egypt's most sacred birds. The spilling of falcon blood was avoided and forbidden, even in magic: its blood sacrifice is denounced as an impious action in *PGM* IV 2593 (= 2656) and, at the moment when the sacrifice of a falcon is required (*PGM* I 1–15), this is carried out by "deification," i.e. by drowning the animal without spilling its blood. It is unlikely, therefore, that actual blood of falcon was used as ink by Greco-Egyptian magicians.[70] The same goes for the "blood of eel," another unique occurrence in the Greek and Demotic magical papyri, as this animal was also sacred in Egypt.[71] Since the *snake blood* of INK D was a resin, it could also be the case that these other "bloods" were vegetable or mineral substances. However, in the absence of a clear referent, their nature remains obscure. Any attempt to identify them is complicated by the complex net of sympathetic associations of Greco-Egyptian magic — without forgetting eventual alterations that occurred in the textual transmission of the recipes, as well as the

69 See Christiansen et al., *Chemical characterization* (cit. note 57), p. 209.

70 Recent research on this substance confirms this assumption and demonstrates that *hierakeion* (ἱερακεῖον) was actually the name of a plant. Accordingly, *haima* refers here to another sap; see Blanco Cesteros, *Los Hermeneumata* (cit. note 62), pp. 162–165.

71 See, e.g. *Hdt.*, 2.72.3. The sacred nature of this animal has been confirmed by the discovery of mummified eels, placed in small bronze cases, see Karol Myśliwiec, *The Twilight of Ancient Egypt: First Millennium B.C.E.* (Ithaca-London: Cornell University Press 2000), p. 9.

possibility of a problematic translation of substances' names from other languages. Many red pigments — minium, cinnabar, hematite (red ochre), resins, etc. — were used to produce red inks, as indicated by both ancient sources and modern chemical analysis of ancient inks.[72] If this were the case, the "blood of the heart of a slaughtered donkey" (INK B) could be minium. The fact that the blood has to come from "the heart of a slaughtered animal" should not be considered as an argument in favour of its interpretations as actual blood. It could be a later specification introduced by a compiler and/or based on a literal interpretation of the substance name. The loss of awareness about the polysemy of given terms resulted in, for example, Akkadian common botanical names like "dog tongue" or "lion fat" being interpreted as actual animal substances in later herbal treatises and pharmacological traditions.[73]

If so, an ink made with minium and copper would have had a beautiful, metallic red colour like the ink made from the resin called *snake blood* and gold (INK D). The alchemical tradition offers recipes of a couple more inks with this finishing touch: in addition to the aforementioned *MC* 45 (see above, §3), the ink from *MC* 34 is based on minium and gold powder, while *MC* 40 uses ochre and mercury for a similar preparation.

7 Concluding Remarks: "Blood" as a Polysemic Reality in Magic

The magical pantry is characterized by the peculiarity and oddity of its ingredients, something also attested to in the composition of the inks described in Greco-Egyptian magical papyri. In particular, the preference for red apparently made blood the most common ingredient for writing in this ritual field. However, the use of blood as (or in) ink was not a phenomenon restricted to magic, and not all apparent "bloods" mentioned in magical inks referred to actual blood. Some, in fact, conceal the name of other substances, encoded by the transmission or by the ritual practitioners. As a result, when properly analysed, some magical inks are no different from those referred to by other sources.

The point of this study has not been to decipher the *Decknamen* of the Greek Magical papyri, but rather to raise awareness about the problematic polysemy

72 See Christiansen et al., *Chemical characterization* (cit. note 57), *passim*.

73 Erica Reiner, *Astral Magic in Babylonia* (Philadelphia: The American Philosophical Society, 1995), pp. 32–33; an extended analysis of this topic in Maddalena Rumor, Babylonian Pharmacology in Graeco-Roman Dreckapotheke. With an Edition of Uruanna III 1–143 (138), PhD Diss. (Berlin: Freie Universität Berlin, 2015).

of magical ingredients and the multiple linguistic, ritual and textual paths that lead to this phenomenon. The real nature of substances is hidden in diverse and not always intentional ways. The literal reading, for example, of botanical names by insufficiently trained scribes introduced alterations (misinterpretations, inferences, lexical variants, etc.) in the textual transmission that blurred the actual nature of these substances. Although not all the names for magical substances have a dual meaning, scholars working with magical recipes must be aware of the potential polysemy of substances and not assume that ingredients should be interpreted literally, especially those belonging to the magical *Dreckapotheke*.[74]

The different "bloody" ingredients examined in this study have also served to highlight the method for approaching those substances involved in magical rituals and procedures. Considered individually, the hidden polysemy would have been barely noticeable; only a comparative study (that included magical and non-magical texts) yields hints about the non-literal value of these names. The ritual context itself is a valuable asset in the recognition of polysemic terms. The interpretation of many of these terms, however, remains open as a result of the dynamic textual phenomena and the complex connections of sympathies and antipathies that underlie Greco-Egyptian magic. In any case, the more we understand this ritual field and its sources, the better we are able to decode them.

74 The list of *PGM* XII, for example, is full of "bloods," "fats," "excrements," and "semen."

CHAPTER 3

Ink in Herculaneum: A Survey of Recent Perspectives

Vincenzo Damiani

Abstract

This paper provides a brief survey of the latest research concerning the types and chemical characteristics of the ink used in the Graeco-Latin papyri from Herculaneum. According to the *communis opinio*, the Herculaneum inks are no exception to the widespread use of carbon black ink in antiquity. This position has been recently revised on the basis of studies that use X-Ray Phase Contrast Tomography (XPCT) to show a significant presence of lead in the ink of some fragments. This important discovery allows for the possibility of using lead as a contrast agent in order to distinguish the writing from the support in still rolled *volumina* through exposure to synchrotron light.

Keywords

Herculaneum papyri – X-Ray Phase Contrast Tomography (XPCT) – carbon black ink – lead

1 Ink in Greek and Latin Papyri

According to ancient testimonies, the ink used in Greek papyri was produced from soot (carbon black/charcoal that functioned as pigment) resulting from the combustion of materials such as resin, oil or wood, mixed with a carbohydrate binder, usually gum arabic from acacia trees (at a ratio of 1 *uncia* of gum to 3 *unciae* of carbon)[1] in order to obtain blocks that form a suspension when mixed with water:[2]

---

1 Dioscorides, *De materia medica*, 5, 182: μείγνυνται δὲ πρὸς οὐγγίαν α′ τοῦ κόμμεως οὐγγίαι γ′ λιγνύος ("Three *unciae* of soot are mixed with one *uncia* of gum"). All the translations are mine, if not indicated otherwise.

2 Cf. the contribution by Ira Rabin in this volume, in the section *Carbon inks*. The ink was then sold by weight: see Gertrud Herzog Hauser, "Tinte," in *Realencyclopädie der Classischen*

inde collecta (scil. *fuligo*) *partim componitur ex gummi subacta ad usum atramenti librarii, reliquum tectores glutinum admiscentes in parietibus utuntur.*

VITRUVIUS, *De architectura*, 7, 10

Then, once the soot has been collected, a part of it is mixed and kneaded with gum to make writing ink; the painters mix the rest with glue to use it on walls.

omne autem atramentum sole perficitur, librarium cumme, tectorium glutine admixto.

PLINIUS, *Naturalis historia*, 35, 25, 43

All black pigments require exposure to the sun; ink for writing is obtained by mixing with gum, ink for painting by mixing with glue.

The ink (τὸ μέλας, *atramentum*)[3] was prepared in advance and made liquid when needed.[4] It was usually kept in two containers (one for black, one for red ink) that together formed the inkpot (μελανοδοχεῖον, *atramentarium*).[5] When the writing instrument (κάλαμος, *calamus*) was dipped into the ink, the liquid accumulated in the fissure at one end of it.

Carbon-based ink is particularly resistant — as it is chemically inert[6] — but can easily be erased by scraping.[7] There seems to be documentary evidence that already in antiquity metallic ink was used in some cases: it was produced from

Altertumswissenschaft, 1940, Suppl. 7: 1574–1579, p. 1576; Viktor Gardthausen, *Das Buchwesen im Altertum und im byzantinischen Mittelalter* (Leipzig: Veit & Comp., 1911), p. 193 and 203; the source is Diocletian's edict *De pretiis rerum venalium*, 19,11; see Siegfried Lauffer, *Diokletians Preisedikt* (Berlin: De Gruyter, 1971, p. 151).

3 Both terms can refer to different types of dark pigment, i.e. not only to black ink, cf. Herzog Hauser, *Tinte* (cit. note 2), pp. 1574–1575.

4 Cf. Demosthenes, *De corona*, 258: τὸ μέλαν τρίβων, with reference to the action of grinding coal; Galenus, *De simplicium medicamentorum temperamentis ac facultatibus*, vol. 12 p. 226 Kühn: Μέλαν ᾧ γράφομεν, ἱκανῶς καὶ τοῦτο ξηραίνει (scil. ἕλκη) λυόμενον ὕδατι ('if previously solved in water, black ink can be used to dry up ulcers'); cf. also Gardthausen, *Buchwesen*, p. 203; Javier Alonso, Rafael Sabio González, José Manuel Jerez Linde, "Tinteros de bronce romanos de Augusta Emerita," *Archivo Español de Arqueología*, 2019, 92:251–269, p. 253.

5 Mario Capasso, *Introduzione alla papirologia* (Bologna: Il Mulino, 2005), p. 109.

6 Fredrik C. Störmer, I. Lorentzen, Brynjulf Fosse, Mario Capasso, Knut Kleve, "Ink in Herculaneum," *Cronache Ercolanesi*, 1990, 20:183–184, p. 183; Mario Capasso, *Manuale di papirologia ercolanese* (Galatina: Congedo Editore, 1991), p. 222.

7 See Monique Zerdoun Bat-Yehouda, *Les encres noires au Moyen Âge* (Paris: Éditions du CNRS, 1983), p. 15.

a mixture of metallic salts (iron or copper sulphate) with tannin extracted from oak galls. The oldest evidence dates back to the 3rd century.[8] Notwithstanding, it took until parchment was established as the main writing support in the 5th century for its use to spread widely.[9] Mixed carbon-based inks with the addition of metallic components such as iron, copper or lead have also been documented.[10] The use of metallic components in the manufacture of ink entails this becoming much more aggressive when used on papyrus than ink based on carbon black.[11] Moreover, when used on parchment, such type of ink adheres better to the writing surface.[12]

The treatment of papyrus fibers with binding substances containing milk, casein, egg white, gum arabic and starch prevented dispersal of the ink on the

8 See Zerdoun, *Encres* (cit. note 7), pp. 16–17 and 91–92. See also Dioscorides, *De materia medica*, 162: δεῖ δὲ τῆς μὲν ἀσβόλης μνᾶν μίαν λαμβάνειν, κόμμεως δὲ λίτραν μίαν ἡμίσειαν, ταυροκόλλης οὐγγίαν μίαν ἡμίσειαν, χαλκάνθου οὐγγίαν μίαν ἡμίσειαν ('You have to take one *mna* of soot, one and a half *litre* of gum, one and a half *uncia* of glue made from bulls' hides, and one and a half *uncia* of copper sulphate'). On ancient accounts concerning the composition of vitriol, see Vladimir Karpenko, "Vitriol in the History of Chemistry," *Chemicke Listy*, 2002, 96(12):998–1005, p. 998. An updated overview of the different types of ink in Antiquity based on physico-chemical analysis is offered in Ira Rabin, Myriam Krutzsch, "The Writing Surface Papyrus and its Materials. 1. Can the writing material papyrus tell us where it was produced? 2. Material study of the inks," in *Proceedings of the 28th International Congress of Papyrology, Barcelona 2016*, edited by Alberto Nodar and Sofía Torallas Tovar (Barcelona: Publicaciones de l'Abadia de Montserrat, 2019), pp. 773–781, pp. 776–779.

9 Adam Bülow-Jakobsen, "Writing Materials in the Ancient World," in *The Oxford Handbook of Papyrology*, edited by Roger Bagnall (Oxford: Oxford University Press, 2009), pp. 3–29 (p. 18); Thomas Christiansen, "Manufacture of Black Ink in the Ancient Mediterranean," *Bulletin of the American Society of Papyrologists*, 2017, 54:167–195, p. 188.

10 Christiansen, *Manufacture* (cit. note 9), p. 169. P.Leid. V (= P.Leid I 384 verso= PGM/PDM XII = TM 55954) preserved in the Leiden National Museum of Antiquities (Rijksmuseum van Oudheden), and dated to the 3rd century AD, contains the complete recipe of a metallogallic ink (made of metallic salt, oak gall, and gum) with the addition of two additives (myrrh and truffle) [Greek text in: Karl Preisendanz, *Papyri Graecae Magicae. Die griechischen Zauberpapyri*, vol. 2 (Leipzig, Berlin: Teubner, 1931), XII, p. 83]: ζμύρνης δραχμὴ α', μίσυος δραχμαὶ δ', χαλκάνθου δραχμαὶ β', κηκίδων δραχμαὶ β', κόμεως δραχμαὶ γ' ('1 drachma of myrrh, 4 darachmae of truffle, 2 drachmae of copper sulphate, 2 drachmae of oak gall, 3 drachmae of gum'). There are testimonies on a pigment made from cuttlefish ink (Herzog Hauser, *Tinte* (cit. note 2), p. 1577; Zerdoun, *Encres* (cit. note 7), pp. 90–91). The coloured inks were obtained from mineral pigments, such as hematite (iron oxide) in the case of red ink: see Capasso, *Introduzione* (cit. note 5), p. 110; Richard Parkinson, Stephen Quirke, *Papyrus* (London: British Museum Press, 1995), p. 44. Ovidius, *Ars amatoria*, 3, 627–630, informs us about the production of 'invisible' inks.

11 Bülow-Jakobsen, *Writing Materials* (cit. note 9), p. 18.

12 See Zerdoun, *Encres* (cit. note 7), pp. 19–20.

surface during the writing process.[13] Ink absorption could be minimized by polishing the papyrus to smoothen the writing surface; excessively porous fibers, on the other hand, could lead to the absorption of too much ink and, consequently, to the formation of stains, sometimes even forcing the scribe to cut off the affected part of the roll and re-glue the remaining sections.[14] In the case of palimpsest papyri, the ink was washed away by treating the papyrus with a wet sponge and, probably, by rubbing it — in order to reuse the roll without writing on the *verso*.[15]

2 The Herculaneum Collection

The Herculaneum scrolls were found between 1752 and 1754 during an excavation campaign commissioned by King Charles VII of Bourbon. The excavations began in 1750 and led to the development of a system of underground passages where the diggers also came across a building later hypothetically identified as a suburban residence (known as *Villa dei Papiri*) once owned by L. Calpurnius Piso Caesoninus, father-in-law of Julius Caesar.[16] At first glance,

13 Capasso, *Introduzione* (cit. note 5), p. 76. On the contrary, no treatment with other substances seems to have been necessary to ensure that the papyrus fibers adhere to each other: see Bridget Leach, John Tait, "Papyrus," in *Ancient Egyptian Materials and Technology*, edited by Paul T. Nicholson and Ian Shaw (Cambridge: CUP, 2000), pp. 227–253 (pp. 233–234); Tiziano Dorandi, "*praeparatur ex eo charta*. Per una rilettura del capitolo di Plinio (Nat. Hist. XIII 71–83) sulla fabbricazione della carta di papiro," *Zeitschrift für Papyrologie und Epigraphik*, 2017, 202:84. I would like to thank Miriam Blanco and Matteo Martelli for bringing these references to my attention.

14 Capasso, *Introduzione* (cit. note 5), p. 77. Cf. Zerdoun, *Encres* (cit. note 7), p. 14.

15 Parkinson, Quirke, *Papyrus* (cit. note 10), p. 47; Zerdoun, *Encres* (cit. note 7), pp. 85–88. Cf. Martialis, *Epigrammata*, 4, 10, 5–8: *comitetur Punica librum / spongia: muneribus convenit illa meis. / Non possunt nostros multae, Faustine, liturae / emendare iocos: una litura potest* ('The book must be accompanied by a Punic sponge: it is suitable for my gifts. Not even many erasures, Faustinus, can emend my jokes: but one single can'); Suetonius, *Augustus*, 85: *tragoediam magno impetu exorsus, non succedenti stilo, abolevit quaerentibusque amicis, quidnam Aiax ageret, respondit Aiacem suum in spongiam incubuisse* ('Because his style did not satisfy him, he destroyed the tragedy to which he had dedicated himself with great enthusiasm. To the friends who asked him, how was his *Ajax*, he answered that "his *Ajax* had thrown himself on the sponge"'). There is evidence of the existence of a solution based on saffron, water, earth, milk and lentisk juice that ensured a perfect erasure of the previous text: see Capasso, *Introduzione* (cit. note 5), p. 92. On the solution to whiten pearls used to prepare a palimpsest (P.Holm. No. 12), cf. Robert Halleux, *Les alchimistes Grecs* (Paris: Belles Lettres, 2010), p. 114.

16 For an overview of the question, see Mario Capasso, "Who Lived in the Villa of the Papyri at Herculaneum — A Settled Question?," in *The Villa of the Papyri at Herculaneum*, edited by Mantha Zarmakoupi (Berlin/New York: De Gruyter, 2010), pp. 89–113; Kenneth

the papyri preserved in the house were mistaken for carbonised wooden pieces or fishing (or hunting) nets,[17] hence a number of them was discarded as material of little significance. Once recognised for what they actually were, the scrolls, still closed and heavily warped by pyroclastic material (ca. 300–350 °C) that had fallen upon the area during the eruption of Mount Vesuvius in 79 AD,[18] were kept in the Museo Ercolanese in Portici (1758) and later moved to the Museo Archeologico Nazionale in Naples (1806). A first, highly invasive method of unrolling them, known as *scorzatura*,[19] involved vertically cutting the closed roll into two hemi-cylinders and progressively scraping off the layers of each of them, starting from the innermost layer and cutting one's way to the external layer, in order to reach each written surface. The text was copied by professional draftsmen, then the upper layer was scraped again in order to expose the underlying layer and so on. This technique, the consequence of which was the destruction of the main part of the papyrus roll (except for the outer layers, which are still preserved today and named *scorze*, 'peels'), was progressively replaced from 1754 onwards by a machine invented by the Genoese priest Antonio Piaggio. The new mechanism allowed for the unrolling of the scrolls while preserving most of the original layers. The outer layer was slowly detached from the underlying one, reinforced with goldbeater's skin, then cut and fixed on a paper sheet.[20] In 1800 the Prince of Wales (the future king George IV) obtained from Ferdinand IV of Bourbon the authorization to send to Naples John Hayter, his personal chaplain who had studied Classics in Eton and Cambridge, in order to speed up the unrolling, transcription and publication of the papyri. From 1802, Hayter was in charge of coordinating the work on the papyri.[21] The series of pencil facsimiles now stored in the Bodleian Library (Oxford) was mainly issued in this period. A first comprehensive edition of Herculaneum texts was published with the title *Herculanensium voluminum quae supersunt collectio prior* (1793–1855, in 11 volumes) and contained the Greek text along with a Latin translation and a commentary. The second part of the *collectio* (*Collectio altera*) followed a decade later (1862–1876, also in

Lapatin (ed.), *Buried by Vesuvius. The Villa dei Papiri at Herculaneum* (Los Angeles: Getty Publications, 2019); Francesca Longo Auricchio, Giovanni Indelli, Giuliana Leone, Gianluca Del Mastro, *La villa dei Papiri. Una residenza antica e la sua biblioteca* (Roma: Carocci, 2020).

17 Capasso, *Manuale* (cit. note 6), p. 68.

18 On the precise date of the eruption, see Mario Capasso, "La biblioteca di Ercolano. Cronologia, formazione, diffusione," *Papyrologica Lupiensia*, 2017, 26:42–68, p. 44.

19 Capasso, *Manuale* (cit. note 6), p. 89.

20 See now Sofia Maresca, "Early Attempts to Open and Read the Papyri: 1750s–1990s," in Lapatin, *Buried by Vesuvius* (cit. note 16), pp. 28–36; Longo Auricchio, Indelli, Leone, Del Mastro, *Villa* (cit. note 16), pp. 59–64.

21 Capasso, *Manuale* (cit. note 6), p. 100 n. 53.

11 volumes): it only displayed engravings of new texts based on pencil facsimiles (the so-called *Neapolitan* drawings). In 1914, after a long and difficult preparation, a third part appeared, edited by Domenico Bassi (*Collectio tertia*). Bassi, who in 1906 was appointed director of the *Officina dei Papiri Ercolanesi*, the institution that is still in charge of the preservation of the scrolls (the *Officina* is now hosted in the *Biblioteca Nazionale di Napoli*; until 1928 it was part of the *Museo Archeologico Nazionale*) undertook a new cataloguing of the papyri. Subsequently, many of them were placed in special boxes (*cornici*, 'frames') for the first time. Meanwhile, the scholarly work on the texts continued. Between the end of the 19th and the first half of the 20th century, important critical editions were published in Germany and in Italy. A systematic publication of the papyri based on modern ecdotic criteria has been undertaken at the beginning of the 1970s, when Prof. Marcello Gigante established a chair of Herculaneum papyrology at the University of Naples and founded the *Centro Internazionale per lo Studio dei Papiri Ercolanesi* (CISPE).

The Herculaneum scrolls have an inestimable value not only with regard to the history of books and writing in antiquity, but also as witnesses to a period in the history of ancient thought that often lacks direct sources, i.e. that between the 3rd and the 1st century BC. In particular, we would know little of the development of the Epicurean Garden (Κῆπος), a major philosophical school of the Hellenistic era, without the essential contribution of the texts found in Herculaneum. It is still a matter of discussion how the library of the *Villa* had come to contain the selection of books it held at the moment of the eruption.[22] We know that a number of the books were brought to Italy by Philodemus of Gadara.[23] Philodemus was a pupil of Zeno of Sidon, who led the Epicurean school in Athens from circa 100 until 75 BC. Following Zeno's death, Philodemus went to Italy and probably took up residence in the *Villa*. The findings of the last forty years confirm that the library contained volumes belonging to different editions of Epicurus' main treatise Περὶ φύσεως (*On nature*). This bulky work in 37 books, probably written between the end of the 4th and the beginning of the 3rd century BC, is a complex and challenging testimony of Epicurus' teaching and research, which survived almost exclusively in the Herculaneum papyri. Modern scholarship has been able to retrieve about a

---

22 On this issue, see Tiziano Dorandi, "La nuova cronologia della 'Villa dei Papiri' a Ercolano e le sorti della biblioteca di Filodemo," *Würzburger Jahrbücher für die Altertumswissenschaft*, 2017, 41:181–203; Capasso, *Biblioteca* (cit. note 18); Longo Auricchio, Indelli, Leone, Del Mastro, *Villa* (cit. note 16), pp. 137–191.

23 See Francesca Longo Auricchio, Giovanni Indelli, Gianluca Del Mastro, "Philodème de Gadara," in *Dictionnaire des philosophes antiques*, edited by Richard Goulet (Paris: Éditions du CNRS, 2011), vol. 5, pp. 334–359.

third of the treatise, albeit in an often-fragmentary state.[24] But the core of the library, as it could be hitherto reconstructed, consists of Philodemus' own works. As far as we can surmise, based on the extant material, Philodemus' interests focused on the history of philosophy and biography, ethics, theology, music, rhetoric, poetics and epistemology. The Epicurean Demetrius Laco — a contemporary of Zeno of Sidon — is also well represented in the Herculaneum collection: his extant output concerns both philological and philosophical issues. Among the other Epicureans whose writings have been fragmentarily preserved in Herculaneum, Metrodorus, Hermarchus, Polyaenus, Carneiscus, Colotes, Idomeneus and Polystratus should be mentioned as well.[25] A smaller group of texts represents Stoic writers, such as Chrysippus, and Latin authors.[26]

3 Ink in Herculaneum

The visibility of the ink on the carbonised papyri must have been much better immediately after their discovery than it is now. In his *Sendschreiben von den Herculanischen Entdeckungen* (Dresden, 1762), Winckelmann writes: (p. 83): "The ink of the ancients was not as fluid as ours, and it was not made with vitriol. This can be seen from the colour of the letters, which is even darker than the carbonised papyrus: this makes reading much easier."[27] Despite the particular resistance of carbon-based ink, however, in some cases the carbonisation process that took place during the eruption, together with chemical reactions over the centuries until their discovery, has resulted in the total or partial disappearance of the ink from the writing support.[28]

24 See Giuliana Leone, "Osservazioni sui papiri ercolanesi di Epicuro," *Studi di Egittologia e Papirologia*, 2014, 11:83–109; Tiziano Dorandi, "Modi e modelli di trasmissione dell'opera *Sulla Natura* di Epicuro," in *Questioni Epicuree*, edited by Dino De Sanctis, Emidio Spinelli, Mauro Tulli, and Francesco Verde (Sankt Augustin: Academia Verlag, 2015), pp. 15–52.

25 See Michael Erler, "Die Schule Epikurs," in *Grundriss der Geschichte der Philosophie. Die Philosophie der Antike*, IV.1, edited by Hellmut Flashar (Basel: Schwabe, 1994), pp. 203–362.

26 On the Latin papyri from Herculaneum, see Mario Capasso, *Les papyrus latins d'Herculanum: Découverte, consistance, contenu* (Liège: CEDOPAL, 2011); further Capasso, *Biblioteca* (cit. note 18), pp. 64–68.

27 "Die Tinte der Alten war nicht so flueßig, wie die unsrige, und war nicht mit Vitriol gemacht. Dieses kann erstlich aus der Farbe der Buchstaben geurtheilt werden, welche schwärzer noch, als die gleichsam in Kohlen verwandelten Schriften sind, wodurch das Lesen derselben sehr erleichtert wird."

28 See Capasso, *Biblioteca* (cit. note 18), p. 45. Cf. Francesco Sbordone, "Recenti tentativi di svolgimento dei papiri ercolanesi," *Cronache Ercolanesi*, 1971, 1:23–39, p. 25, reporting the opinion of the Austrian restorer Anton Fackelmann: "Although the rolls have a length of 150–180 cm, only scanty remains of letters appear in a few places. Unfortunately, the

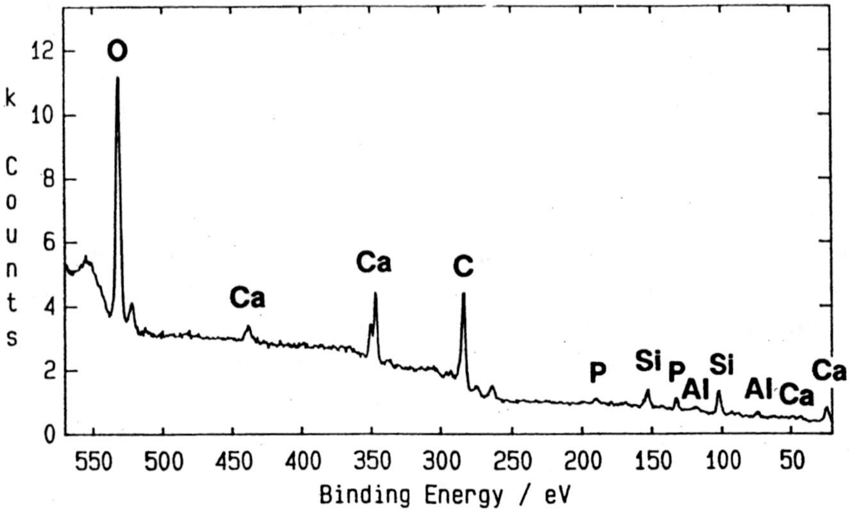

FIGURE 3.1 Energy spectrum of the surface of a papyrus of Herculaneum, obtained through the use of an electron microscope and an energy-dispersive X-ray analyzer

Experiments conducted between the end of the 1980s and the beginning of the 1990s concerning the ink composition of some fragments of the Herculaneum collection confirm that the scrolls essentially correspond with the Greek papyri found in the sands of Egypt, since their ink is primarily carbon-based (Fig. 3.1).[29]

About a decade ago, an article by Brent Seales[30] outlined the first results of analyses performed on Herculaneum papyri using different imaging methods, from X-ray techniques (EDX, PIXE) to magnetic resonance imaging (MRI) and computed tomography (CT). It established the main criteria to be considered in order to obtain a sufficient contrast between writing and support: 1) the difference in density between ink and papyrus; 2) the morphology of the writing, which in various ways interacts with the support and produces

ancient Egyptian soot ink is much more resistant than the Roman ink. The Egyptian ink is a gum-resin mixture with soot, an ink that can also be placed in boiling water if necessary (mummy cartonnage). In Italy, on the other hand, the ink (according to Pliny) was made with grated charcoal and gum, an ink that adheres more to the surface and tends to form dust."

29 Störmer, Lorentzen et al., *Ink* (cit. note 6), p. 183. For a review of recent applications of experimental sciences to the Herculaneum Papyri see Gianluca Del Mastro, "Papyrology and Experimental Sciences," *Research Trends In Humanities Education & Philosophy*, 2019, 6:8–15.

30 W. Brent Seales, "Lire sans détruire les papyrus carbonisés d'Herculanum," *Comptes rendus des séances de l'Académie des Inscriptions et Belles-Lettres*, 2009, 153(2):907–923.

specific structural features; and 3) the chemical-physical analysis of the elements that constitute the ink and the papyrus. The analyses carried out on some fragments resulted in the identification of calcium (Ca) as a possible distinctive element between ink and papyrus (with the concentration of calcium (Ca) being correlated to the areas where the ink is present)[31] and the detection of lead (Pb) and strontium (Sr) (which was not found during the experiments by Störmer, Lorentzen et al.) in the ink's composition.[32]

In any case, it has still proven impossible to distinguish the writing of unwound papyrus rolls from carbonised fibres using conventional X-ray techniques based on the different level of absorption of the rays to which the object is exposed.[33] The study of the ink composition in Herculaneum has therefore been the subject of new research in recent years. Interesting results have been achieved through the introduction of innovative screening methods, such as X-Ray Phase Contrast Tomography (XPCT), a technology sensitive to materials with similar characteristics, as in the case of the carbon-based ink used for writing and the carbonised surface of the papyri (both of which exhibit weak levels of absorption).[34] The possibility of distinguishing the writing from the surface presupposes the identification of the precise chemical composition of the ink, which determines the choice of the optimal X-ray wavelength, in order to achieve sufficient contrast to make the writing legible.[35] Another physical feature that helps to isolate the ink is the fact that it usually shows a slight relief from the surface of the papyrus (the carbon black ink does not completely penetrate the fibres), causing the X-rays to undergo a minimum deviation at that point.[36]

31 Seales, *Papyrus carbonisés* (cit. note 30), p. 914; cf. W. Brent Seales, Daniel Delattre, "Virtual Unrolling of Carbonized Herculaneum Scrolls: Research Status (2007–2012)," *Cronache Ercolanesi*, 2013, 43:191–208, p. 196.

32 Seales, *Papyrus carbonisés* (cit. note 30), p. 917; cf. Seales, Delattre, *Virtual Unrolling* (cit. note 31), pp. 197–198.

33 Ana S. Leal, Silvia Romano, Vito Mocella, "Ink Study of Herculaneum Papyri," *Manuscript Cultures*, 2018, 11:17–20, p. 17; but see also Clifford Seth Parker, Stephen Parsons, Jack Bandy *et al.*, "From invisibility to readability: Recovering the ink of Herculaneum," *PLoS ONE* 14(5): e0215775 [DOI: 10.1371/journal.pone.0215775].

34 Gianluca Del Mastro, Daniel Delattre, Vito Mocella, "Una nuova tecnologia per la lettura non invasiva dei papiri ercolanesi," *Cronache ercolanesi*, 2015, 45: 227–230, p. 229; Vito Mocella, Emmanuel Brun, Claudio Ferrero, Daniel Delattre, "Revealing Letters in Rolled Herculaneum Papyri by X-Ray Phase-contrast Imaging," *Nature Communications*, 2015, 6 (5895):2–6, p. 2.

35 Leal, Romano, Mocella, *Ink* (cit. note 33), pp. 17, 18; Del Mastro, Delattre, Mocella, *Tecnologia* (cit. note 34), p. 228.

36 Del Mastro, Delattre, Mocella, *Tecnologia* (cit. note 34), pp. 228, 230; Mocella, Brun et al., *Revealing Letters* (cit. note 34), p. 3. Cf. Seales, *Papyrus carbonisés* (cit. note 30), p. 912.

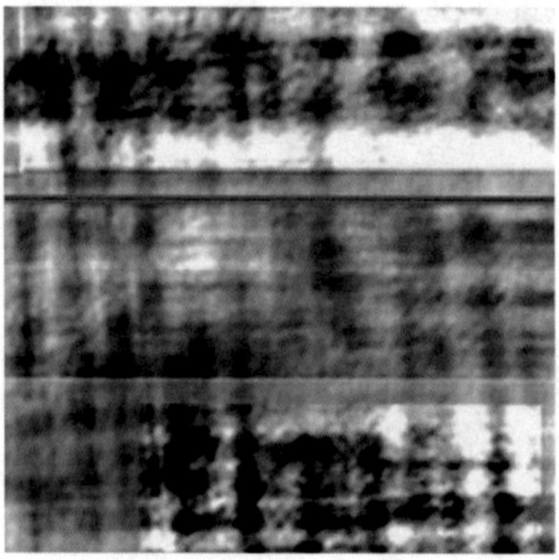

FIGURE 3.2
Results of the scan of *PHerc.Paris.* 1

Initial experiments with the application of XPCT on Herculaneum papyri were conducted in December 2013 by a team led by the Italian physicist Vito Mocella at the ESRF (European Synchrotron Research Facility) in Grenoble, France.[37] There, two rolls from the collection of the Institut de France, donated to Napoleon Bonaparte by Ferdinand IV of Bourbon in 1802,[38] (*PHerc.Paris.* 4, rolled; *PHerc.Paris.* 1, a fragment with several layers of papyrus superimposed on each other) were subjected to radiation from one of the light lines produced by the synchrotron through the magnetically induced deviation of accelerated electrons (ID 17). The diffraction of the X-ray beam, which was then processed into three-dimensional images, revealed short sequences of letters. For *PHerc.Paris.* 1 it was possible to identify the verbal forms ΠΙΠΤΟΙΕ (presumably from πίπτω, "to fall") and ΕΙΠΟΙ (perhaps from a *verbum dicendi*) (Fig. 3.2).[39]

The scan of *PHerc.Paris.* 4, which has a very complex morphology due to it being exposed, at the time of the eruption, to considerable pressure that has greatly deformed its structure, revealed some groups of letters, which can be interpreted, with due caution, as ΑΡΝ, ΗΕΥ and ΚΙ (Fig. 3.3).

Similar results were obtained in a series of experiments conducted simultaneously and independently by another team of scientists, who combined XPCT with three-dimensional computation algorithms — known as "virtual unrolling" — adapted to the particular structural deformations of the

37 A previous synchrotron light experiment was conducted at the Stanford Linear Accelerator without any decisive results in ink detection: see Seales, Delattre, *Virtual Unrolling* (cit. note 31), p. 206.

38 Cf. Marcello Gigante, "I papiri ercolanesi e la Francia," in *Contributi alla storia della Officina dei Papiri Ercolanesi* 2, a cura di Marcello Gigante (Roma: Istituto Poligrafico e Zecca dello Stato, 1986), pp. 27–35.

39 Mocella, Brun et al., *Revealing Letters* (cit. note 34), p. 3.

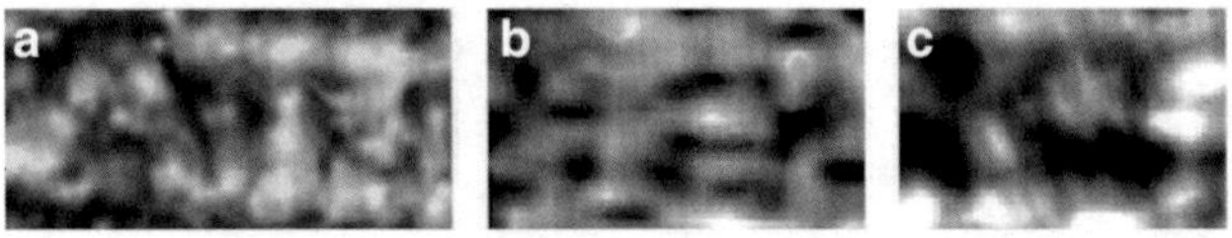

FIGURE 3.3 Results of the scan of *PHerc.Paris.* 4

Herculaneum papyri. They also managed to make some portions of text from still rolled scrolls visible (*PHerc.* 375 and *PHerc.* 495).[40]

Recent analyses carried out on some fragments of Herculaneum papyri by synchrotron X-ray-based techniques have shown the presence of lead (in a state similar to lead (II) acetate) within the ink — a surprising result that challenges the *communis opinio* on the exclusively carbon-based composition of Herculaneum ink. Hypotheses have been made regarding the causes of this phenomenon, which may be due to:[41]

a) Contamination of the water used to fabricate the inks.
b) The presence of lead in the bronze containers in which the ink was stored.
c) An intentional use, either as an additional pigment or as a binding medium to accelerate the drying process.[42]

According to Tack, Cotte et al. (cit. note 41), the concentration of lead detected in the ink of the fragments used as test samples is too high (11–21 µg/cm2 in

40 Inna Bukreeva, Alberto Mittone, Alberto Bravin, Giulia Festa, Michele Alessandrelli, Paola Coan, Vincenzo Formoso et al., "Virtual Unrolling and Deciphering of Herculaneum Papyri by X-Ray Phase-Contrast Tomography," *Scientific Reports*, 2016, 6 (27227):2–6. See also, more recently, Inna Bukreeva, Graziano Ranocchia, Vincenzo Formoso, Michele Alessandrelli, Michela Fratini, Lorenzo Massimi, Alessia Cedola, "Investigation of Herculaneum Papyri by X-Ray Phase-Contrast Tomography," in *Nanotechnologies and Nanomaterials for Diagnostic, Conservation and Restoration of Cultural Heritage. A Volume in Advanced Nanomaterials*, edited by Giuseppe Lazzara and Rawil Fakhrullin (Amsterdam [etc.]: Elsevier, 2019), pp. 299–324, which integrates the results of the previous study with an extensive report on the material characteristics (such as the arrangement of whorls after the mechanical action of the pyroclastic material, the damage occurred at the time of the eruption and during cataloguing and storage, the presence of foreign bodies such as sand and pebbles) of the rolled volumes subjected to XPCT. On the first virtual unrolling experiments conducted on the papyri of Herculaneum see Seales, Delattre, *Virtual Unrolling* (cit. note 31).

41 See Pieter Tack, Marine Cotte, Stephen Bauters, Emmanuel Brun, Dipanjan Banerjee, Wim Bras, Claudio Ferrero et al., "Tracking Ink Composition on Herculaneum Papyrus Scrolls: Quantification and Speciation of Lead by X-Ray Based Techniques and Monte Carlo Simulations," *Scientific Reports*, 2016, 6 (20763):1–7, p. 2; Lars Krutak, "Ink," in *The Encyclopedia of Archaeological Sciences*, edited by Sandra L. López Varela (Malden, MA: Wiley-Blackwell, 2019), pp. 942–945, p. 943.

42 Cf. Zerdoun, *Encres* (cit. note 7), p. 16 on the influence of additives on ink properties.

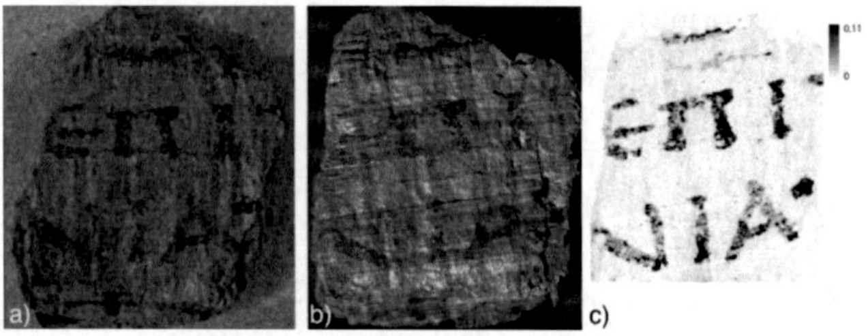

FIGURE 3.4 Comparison of natural light images (a), infrared images (b) and the lead distribution map obtained by using X-rays

one case, 79–89 µg/cm2 in the other) to be traced back to contamination of the water used as a solvent (the average value in the latter case being less than 1.5 mg/L) or to the storage of the pigments in bronze containers (unlike lead, no co-distribution of copper (Cu) with the writing was detected).[43] The authors of the study consider it more likely that the presence of lead is due to an intentional addition as a pigment or as a drying agent:[44] this would lead to classifying the ink used in some of the papyri as a "mixed type" ink.[45] The importance of this discovery is the possibility of using lead as a contrast agent to bring out the writing on the support. In the study by Brun, Cotte et al., the mapping of the presence of lead in two fragments by synchrotron rays allows for the visualization of groups of letters even more clearly than in natural light or infrared photographs (the latter obtained at a wavelength of 940 nm)[46] (Fig. 3.4).

In comparison with the distribution maps of other elements detected in the composition of the ink, lead seems thus to be the best element for discriminating between ink and writing support.[47] Leal, Romano et al. have

43 Tack, Cotte et al., *Ink Composition* (cit. note 41), p. 3; cf. Emmanuel Brun, Marine Cotte, Jonathan Wright, Marie Ruat, Pieter Tack, Laszlo Vincze, Claudio Ferrero et al., "Revealing Metallic Ink in Herculaneum Papyri," *Proceedings of the National Academy of Sciences of the United States of America*, 2016, 113 (14):3751–3754, p. 3752.

44 Tack, Cotte et al., *Ink Composition* (cit. note 41), p. 5.

45 See. n. 10; Zerdoun, *Encres* (cit. note 7), p. 20: "Ce sont par exemple des *encres au carbone* auxquelles le préparateur a ajouté des extraits aqueux de produits tannants ou des sels métalliques."

46 On infrared images of the Herculaneum papyri, see Steven W. Booras, David R. Seely, "Multispectral Imaging of the Herculaneum Papyri," *Cronache Ercolanesi* 1999, 29:95–100. Recent developments, such as their use in combination with RTI (Reflectance Transformation Imaging), are reported in Kathryn Piquette, "Illuminating the Herculaneum Papyri: Testing New Imaging Techniques on Unrolled Carbonised Manuscript Fragments," *Digital Classics Online*, 2017, 2(1):80–102.

47 Brun, Cotte et al., *Metallic Ink* (cit. note 43), p. 3752. (Leal, Romano et al., *Ink*, p. 19).

recently published the results of an experiment conducted on facsimiles of Herculaneum papyri, in which a fragment of papyrus written with blue ink containing lead (II) acetate was placed inside a roll of papyrus written using commercial Chinese black ink and subsequently scanned by XPCT: the experiment reveals that it is possible to decipher, in the three-dimensional reconstruction of the roll, some letters written on the fragment with ink containing lead.

Nevertheless, claiming that the inks of the Herculaneum papyri *all* contained lead would be incorrect. The absence of this element in the results of the study published in 2015 by Mocella, Brun et al. (cit. note 34) rather suggests that the composition of the inks may have varied according to the chronology of the papyri (which, in the case of the library of Herculaneum, covers a spectrum of at least three centuries, from the 3rd cent. BC to the end of the 1st cent. AD) and the scribes who copied the texts.[48] Still, the results of the aforementioned studies represent an important starting point for the development of more refined techniques capable of better distinguishing the writing without the need to handle extremely fragile artefacts, thus ensuring their preservation.

48 *Ibid.*, pp. 3752–3753. See Zerdoun, *Encres* (cit. note 7), p. 13: "En effet, pour une recette d'encre possédant les mêmes composants de base, mélanger ceux-ci en proportions différentes peut donner une encre de réactivité chimique variable avec, pour conséquence, des effets divers sur le support. Ne serait-ce pas ce qui s'est souvent produit lorsque les scribes, à partir d'une recette donnée, auraient, pour une raison quelconque, changé les proportions, créant ainsi une encre de même nature, mais de réactivité différente ?" On the dating of the Herculaneum papyri see Capasso, *Manuale* (cit. note 6), pp. 151–198 and Capasso, *Biblioteca* (cit. note 18), pp. 47–50.

CHAPTER 4

Material Studies of Historic Inks: Transition from Carbon to Iron-Gall Inks

Ira Rabin

Abstract

This chapter offers observations and considerations concerning black writing inks encountered in writing supports transmitting documentary and literary texts of the late Antiquity and early Middle Ages. It discusses different types of inks, the methods of their detection and their use in different times and geographical areas.

Keywords

writing inks – ink fingerprint – material analysis

The transition period from carbon ink, commonly associated with Antiquity, to iron-gall ink, extremely popular in the medieval Middle East and Europe, provides the focus for our investigative work. Our studies combine a search of extant written records with an experimental determination of the ink composition in selected, preferably dated and localized manuscripts.[1] Though

1 Claudia Colini, From Recipes to Material Analysis: the Arabic Tradition of Black Inks and Paper Coatings (9th to 20th century), (PhD Diss., Hamburg University, 2018); Zina Cohen, Composition Analysis of Writing Materials in Geniza Fragments, (PhD Diss., BAM, EPHE and Hamburg University, 2020); Tea Ghigo, Systematic Scientific Approach to the Study of Inks from Coptic Manuscripts, (PhD Diss., BAM, Sapienza Rome and Hamburg University, 2020). Much of the research presented here is carried out in the context of the SFB 950 'Manuskriptkulturen in Asien, Afrika und Europa,' funded by the German Research Foundation (Deutsche Forschungsgemeinschaft, DFG) and within the scope of the Centre for the Study of Manuscript Cultures (CSMC). I would like to acknowledge the help of the personnel of the museums, archives and libraries who grant access to manuscripts and support this work. My thanks go to Oliver Hahn, the head of the BAM 4.5 division, for his contribution and other department members at the BAM and CSMC who participate in this research: Olivier Bonnerot, Sebastian Bosch, Zina Cohen, Claudia Colini, Ines Feldmann, Tea Ghigo and Ivan Shevchuk. My special thanks go to my husband, Marcello Binetti, for fruitful discussions and proof-reading of the manuscript.

ink studies have become increasingly popular in the last five to ten years, no large-scale research in this field has yet been conducted and, therefore, a general picture cannot be drawn at this stage.

To outline the inks of the transition period, we begin with the definitions of known inks, their properties and the techniques that are used to identify and study them.

1 Carbon Inks

One of the oldest writing and drawing pigments is produced by mixing soot or charcoal with a binder dissolved in a water soluble medium. Thus, along with soot, binders such as gum arabic (ancient Egypt) or animal glue (China) form the main components of soot inks. These inks were usually stored as dry cakes. To produce a writing liquid, they were mixed with water directly before an inscription was planned. Carbon inks form a suspension and, therefore, do not penetrate a substrate (papyrus, parchment or sized paper), but are well localized on the surface. Traditionally, one identified carbon inks using infrared photography, since their deep black colour does not change under infrared light. In recent decades, Raman spectroscopy has become more popular in the identification of carbon inks and it was hoped that this method would distinguish between charcoal and soot as precursors for the ink. Unfortunately, this is not yet the case. Similarly, the claim that Raman spectroscopy can be used as a non-invasive dating tool remains unsubstantiated.[2] There is a number of extant recipes, ranging from the late Antiquity[3] to the Middle Ages.[4] Carbon inks containing a considerable amount of copper were detected on a small number of the Dead Sea Scrolls.[5] Recently, carbon inks containing significant

2 Sarah Goler, James T. Yardley, Angela Cacciola, Alexis Hagadorn, David Ratzan, Roger Bagnall, "Characterizing the Age of Ancient Egyptian Manuscripts though Micro-Raman Spectroscopy," *Journal of Raman Spectroscopy*, 2016, 47:1185–93.

3 Dioscorides, *De materia medica. Codex Neapolitanus, Napoli, Biblioteca nazionale, Ms. Ex. Vindob. Gr. 1*, kommentiert von Carlo Bertelli, Salvatore Lilla, Guglielmo Cavallo, 2 vols, Rom: Salerno Editrice / Graz: Akademische Druck- und Verlagsanstalt, 1988/1992; Pliny, *The Natural History*: in 10 volumes, vol. 9: Books XXXIII–XXXV, ed. and transl. by Harris Rackham, Cambridge, MA: Harvard University Press 1995 (The Loeb Classical Library 394; reprint).

4 Armin Schopen, *Tinten und Tuschen des arabisch-islamischen Mittelalters*, (Göttingen: Vandenhoeck & Ruprecht, 2014).

5 Yoram Nir-El, and Magen Broshi, "The Black Ink of the Qumran Scrolls," *Dead Sea Discoveries*, 1996, 3:158–167.

quantities of lead (Pb) were identified on a Herculaneum papyrus.[6] Given that, in both cases, carbon constitutes the colorant of the inks we consider these varieties to belong typologically to the class of carbon inks.

2 Plant Inks

Plant or tannin inks are solutions of tannins extracted from various plants and are brown in colour. Plant inks are absorbed by the substrate; the degree of absorption largely depends on the nature of that substrate. They have a characteristic homogeneity and show no crystallization. They gradually fade under light of a growing wavelength and become transparent under light of ca. 750 nm. They are best identified by their physical and optical properties. The most well-known plant ink, at least for the Western Middle Ages, is the thorn or Theophilus' ink,[7] the elaborate recipe for which is recorded in Theophilus' 12th century work *De diversis artibus*.[8] Another recipe is offered by Martianus Capella in his encyclopedic work. Unfortunately, no systematic study of the historical use of these inks has been compiled: their use has been only occasionally reported in different scriptoria. Inks of brown-reddish colour in Near East manuscripts have also been identified as pure vegetable extracts.[9]

3 Iron-Gall Inks

Iron-gall inks dominated the black to brown palette of writing materials in the manuscripts ranging from the Middle Ages to the 19th century. They are produced by the reaction between iron(II) and gallic acid, which results, initially, in a colourless, soluble complex that forms black, water-insoluble pigment upon oxidation. In addition, a great variety of medieval to modern recipes name various water-soluble binders and solvents such as water, wine or vinegar that

6 Emmanuel Brun, Marine Cotte, Jonathan Wright, Marie Ruat, Pieter Tack, Laszlo Vincze, Claudio Ferrero, Daniel Delattre, Vito Mocella, "Revealing Metallic Ink in Herculaneum papyri," 2016, *Proceedings of the National Academy of Sciences of the United States of America*. 113, 201519958. <https://doi.org/10.1073/pnas.1519958113> (last accessed 13th March 2019).

7 Monique Zerdoun Bat-Yehouda, *Les encres noires au Moyen Âge: jusqu'à 1600* (Paris: CNRS, 1983), pp. 156–165 offers a detailed discussion of this ink, which could be a plant ink or an imperfect iron-gall ink if one translates *atramentum* as carbon black or vitriol, respectively.

8 Charles R. Dodwell, *Theophilus, De diversis artibus. Theophilus, the Various Arts. Translated from the Latin with Introduction and Notes* (London: Thomas Nelson, 1961), pp. 34–35.

9 Zerdoun, *Les encres noires* (cit. note 7), p. 117.

were used to extract gallic or tannic acids from gall nuts — diseased formations on the leaf buds, leaves, and fruits of various species of oak, caused when parasitoid wasps deposit their eggs in them — and tree bark. Iron(II) sulfate (also known as 'green vitriol' because of its colour and its glassy appearance) is the most frequently named ingredient in ink formulas. Natural vitriol[10] consists of a mixture of metallic sulfates (iron sulfate, copper sulfate, manganese sulfate, zinc sulfate, etc.) with relative weight contributions characteristic of the source or purification procedure.[11] Since their elemental composition is rich in metals that are easily detected by the X-Ray Fluorescence (XRF) analysis, it has become the method of choice for identification and classification of iron-gall inks. This technique benefits from the availability of a variety of transportable instruments ranging from single spot to high-resolution scanning equipment, as well as from a wealth of knowledge and experience that has been accumulated in the characterization of historical inks via this technique. Specifically, the development and use of the fingerprint model, based on the quantitative and semi-quantitative determination of inorganic ingredients normalized to iron, as a main ingredient of the iron-gall inks, allows their reliable classification.[12]

Microscopically, iron-gall inks appear highly inhomogeneous in colour and texture, with traces of dark crystals. They gradually lose opacity in the near-infrared region and become invisible under infrared light of ca. 1400 nm. Finally, they display a characteristic spectrum when investigated by Raman spectroscopy.[13] Since the inks contain both soluble and insoluble parts, they usually penetrate a substrate such as paper but display a preferential penetration on the flesh side of parchment.

Medieval iron-gall inks, with their distinct metallic fingerprint, also attract a great deal of attention. Recent technological developments in the field of non-destructive testing and growing interdisciplinary collaborations have led

10 Deatiled account of historic use of the vitriol lies beyond the scope of this article. Vladimir Karpenko and John A. Norris (Chem. Listy, 2002, 96. 997–1005) offer detailed account of the terminology and sources of vitriol in Antiquity and Middle Ages.

11 Christoph Krekel, "Chemische Struktur historischer Eisengallustinten," in G. Banik and H. Weber (eds), *Tintenfraßschäden und ihre Behandlung, Werkhefte der staatlichen Archivverwaltung Baden-Württemberg*, Stuttgart: Kohlhammer, pp. 25–36 (Serie A Landesarchivdirektion, 10).

12 Oliver Hahn, "Eisengallustinten — Materialanalyse historischer Schreibmaterialien durch zerstörungsfreie naturwissenschaftliche Untersuchung," *Editio, Internationales Jahrbuch für Editionswissenschaft* 2006, pp. 143–157.

13 Alana Lee, Vincent Otieno-Alego, Dudley C. Creagh, "Identification of Iron-Gall Inks with Near-Infrared Raman Microspectroscopy," *Journal of Raman Spectroscopy*, 2008, 39:1079–1084.

to a broadening of the field of codicology to include experimental characterization of writing materials. In this new field, we work together with papyrologists, codicologists and palaeographers, who include the primary tests of the inks in their studies choosing representative manuscripts for in-depth characterization.[14] Interestingly, in the Middle East, a multitude of various recipes and scarcity of analytic data contrasts strongly with the situation in Europe, where the first recipes do not appear before the 12th century; but a multitude of analytic results clearly demonstrates that by that time the iron-gall inks had already become a dominant ink type all over Europe.

To facilitate the interdisciplinary work, we have developed a three-step standard protocol for the determination of ink composition. Our protocol uses mobile and non-invasive equipment to conduct the studies on site and consists of a primary screening to broadly classify the ink as carbon, plant (tannin) or iron-gall. This is followed by the determination of the elemental composition using X-ray fluorescence spectrometry (XRF) and a subsequent in-depth analysis using vibration spectroscopic techniques: FTIR and Raman. The primary screening is carried out by means of NIR (near infrared radiation) reflectography. As mentioned above, optical differences between carbon, plant (tannin) and iron-gall inks are best recognized by comparing their response to infrared, rather than near-infrared light: carbon ink has a deep black colour, iron-gall ink becomes transparent above 1400 nm and tannin ink disappears at about 750 nm.[15] We have simplified the analysis using a small USB microscope with built-in NIR (940 nm) and UV (395 nm) LED in addition to an external white

14 Ira Rabin, "Building a Bridge from the Dead Sea Scrolls to Mediaeval Hebrew Manuscripts," in *Jewish Manuscript Cultures. New Perspectives*, edited by I. Wandrey, (Berlin/Boston: De Gruyter, 2017), pp. 309–322; Ira Rabin, "Instrumental Analysis in Manuscript Studies in Comparative Oriental Manuscript Studies. An Introduction," edited by A. Bausi et al. (Hamburg: COMSt; Tradition, 2015), pp. 27–30; Denis Nosnitzin and Ira Rabin, "A Fragment of an Ancient Hymnody Manuscript from Mǝʾǝsar Gwǝḥila (Tǝgray, Ethiopia)," *Aethiopica*, 2014, 17:65–77; Ira Rabin, Oliver Hahn, Marcello Binetti, "Inks of Medieval Hebrew Manuscripts: A Typological Study," *Manuscript Cultures*, 2014, 6:119–131; Ira Rabin, Oliver Hahn, and Miriam Geissbühler, "Combining Codicology and X-Ray Spectrometry to Unveil the History of Production of Codex Germanicus 6 (Staats- und Universitätsbibliothek Hamburg)," *Manuscript Cultures*, 2015, 7:126–131; Miriam Geissbühler, Georg Dietz, Oliver Hahn, and Ira Rabin, "Advanced Codicological Studies of Cod. germ. 6: Part 2," *Manuscript Cultures*, 2018, 11:133–140; Marco Heiles, Ira Rabin, and Oliver Hahn, "Palaeography and X-Ray Fluorescence Spectroscopy: Manuscript Production and Censorship of the Fifteenth Century German Manuscript, State and University Library Hamburg, Cod. germ. 1," *Manuscript Cultures*, 2018, 11:109–139.

15 Ralf Mrusek, Robert Fuchs, and Doris Oltrogge, "Spektrale Fenster zur Vergangenheit — Ein neues Reflektographieverfahren zur Untersuchung von Buchmalerei und historischem Schriftgut," *Naturwissenschaften*, 1995, 82:68–79.

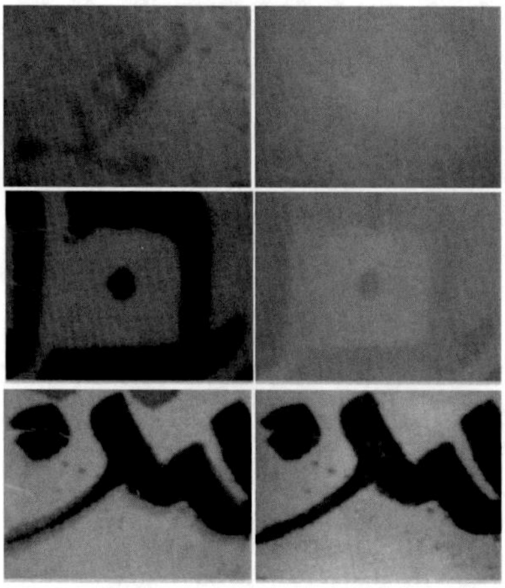

FIGURE 4.1
Visible (left) and NIR (right) images letters written with typologically different inks. Upper row: plant (tannin) ink; middle row: iron-gall ink; bottom row: carbon ink

light source.[16] Comparing the images under white and near-infrared illumination, we determine the ink typology by observing changes in the opacity of the ink. Here, carbon-based inks (Fig. 4.1, bottom row) show no change in their opacity when illuminated with NIR wavelength, while the opacity of iron-gall inks changes considerably (Fig. 4.1, middle row) and plant (tannin) inks become transparent (Fig. 4.1, upper row). Usually, our partners from the Humanities include the first step of the protocol in their manuscript assessment.

The second step of the protocol consists of the XRF analysis to obtain contributions of the inorganic components of the ink under investigation. In the case of iron-gall inks, we establish the fingerprints, i.e. the characteristic ratios of the vitriolic components of the ink (Fig. 4.2). In the case presented in Fig. 4.2 we addressed a specific question: whether composition of the ink would support paleographic attribution of three fragments to a single manuscript. The analysis shows that the fingerprint of the ink, the ratio of copper / iron and zinc / iron is indeed constant in all three fragments, supporting the hypothesis of a single manuscript. In contrast, the ink of the vowel displays a different fingerprint confirming that the vocalization points were not written together with the consonantal text.

In the case of carbon-based inks, XRF analysis serves to discover the presence of metals and identify mixed inks, i.e. inks produced by addition of various metals to the soot inks and intentional mixing of iron-gall and soot-based inks, respectively. Mixed inks have received no or little attention in scholarly

16 Ira Rabin, Roman Schütz, Anka Kohl, Timo Wolff, Roald Tagle, Simone Pentzien, Oliver Hahn, and Stephen Emmel, "Identification and Classification of Historical Writing Inks in Spectroscopy," *Comparative Oriental Manuscript Studies Newsletter*, 2012, 3:26–30.

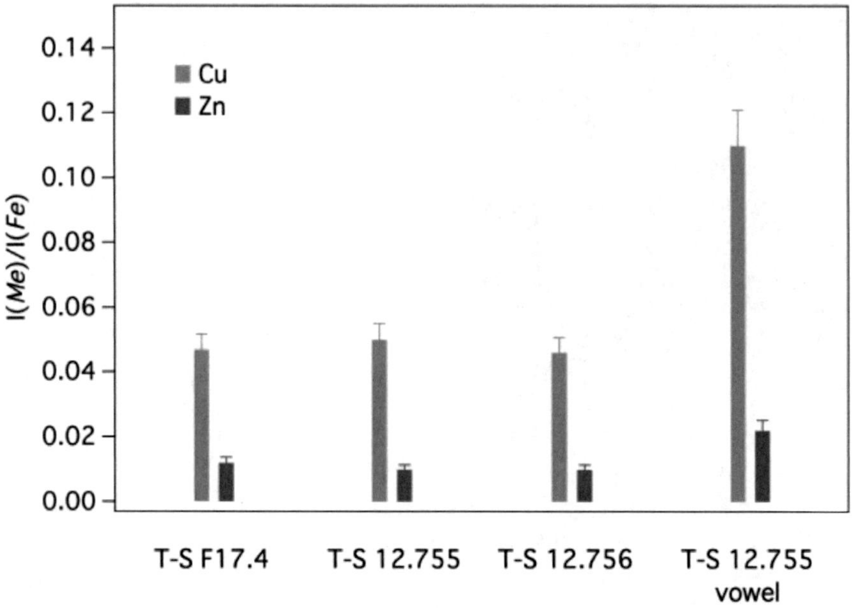

FIGURE 4.2 Comparison of the fingerprint of writing ink in the fragments T-S F17.4, T-S 12.755 and T-S 12.756 (Cambridge University Library). Intensities of the metallic components, copper (Cu, grey) and zinc (Zn, black), are normalized to that of iron (Fe). Here, ink analysis demonstrated that three different fragments belong to the same manuscript

and material studies because their identification and recognition of their importance are recent discoveries. Yet, judging by the Arabic medieval ink recipes, mixed inks played an important role in the Islamicate world.[17] Their detection, and especially their classification, is quite difficult if one is limited to non-invasive methods. In the third step of our protocol, used for these specific cases, we perform FTIR spectroscopy to collect information on the chemical composition of the binders and Raman spectroscopy to determine the co-presence of carbon and iron-gall ink.

The analytic studies of the inks indicate that typologically distinct inks existed as early as the 3rd century BCE.[18] It must be noted that the oldest extant recipe of a predecessor of the iron-gall ink is authored by Philo of Byzantium,

17 Claudia Colini, Oliver Hahn, Olivier Bonnerot, Simon Steger, Zina Cohen, Tea Ghigo, Thomas Christiansen, Marina Bicchieri, Paola Biocca, Myriam Krutzsch, and Ira Rabin, "The Quest for the Mixed Inks," *Manuscript Cultures* 2018, 11:41–48.

18 Elisabeth Delange, Maurice Grange, Bruce Kusko, Eve Menei, "Apparition de l'encre métallogallique en Égypte à partir de la collection de papyrus du Louvre," *Revue d'Égyptologie*, 1990, 41:213–217.

who lived in about the same time.[19] Pliny's detailed account of the manufacture of various soot-based inks in the 1st century CE indicates that, despite its apparent simplicity, the production of high-quality, pure soot was not an easy task in antiquity.[20] His account fits well with the scientific results of the inks analysis: in the period between the 3rd century BCE and the 5th century CE, we find various black writing inks that can be roughly grouped in the following types: pure soot or charcoal-based inks; soot inks with the addition of copper or lead; mixed inks containing soot and iron-gall inks or tannins; and pure iron-gall inks. It is rather tempting to correlate the emergence of such a variety with the growing need for ink and experimentation to substitute rather expensive carbon inks, the main writing ink of antiquity: it could be viewed as a by-product of the expansion of the bureaucracy needed for the proper functioning of the expanding empire, following the conquests of Alexander the Great. On the other hand, finding a great variety of inks might simply reflect the growing number of analyses conducted in recent decades. Christiansen, for instance, suggests that metals in the soot inks appear due to their presence in the metallurgical soot used as a precursor in Egypt.[21] In such a case, we would also expect to find metals in the inks of pharaonic Egypt. However, the lack of reference to metallurgical soot in the chapters dedicated to the inks in the books of Pliny the Elder supports our skepticism about this explanation.

It is unfortunate that the scarcity of known recipes from the period under investigation (i.e. 3rd century BCE and 7th century CE) cannot account for the variety of inks detected analytically. There is an urgent need to reassess the written sources, paying specific attention to the production of writing black inks. Interestingly, all the extant early recipes for metal-containing inks always refer to copper-based substances rather than iron, though only iron ions produce black precipitate upon reaction with galls.[22] The clear distinction between copper and iron sulphates probably marks the beginning of the iron-gall inks era. On one hand, our preliminary results indicate that iron-gall ink seems to be established by the 5th century CE, at least in the Middle East. On the other

19 Zerdun, *Les encres noires*, (cit. note 7), p. 93.

20 Pliny, *Naturalis historia*, Book XXXV <http://www.perseus.tufts.edu/hopper/text?doc=Plin.+Nat.+toc> (last accessed 13th March 2019); see also Pliny, *The Natural History* (cit. note 3).

21 Thomas Christiansen, "Manufacture of Black Ink in the Ancient Mediterranean," *Bulletin of the American Society of Papyrologists* 2017, 54:167–195.

22 Recently mixed inks containing soot inks and a copper based precursor of iron-gall ink were found on a documentary Greek papyrus stored in Florence. See Ira Rabin, Carsten Wintermann, Oliver Hahn, "Ink Characterization, Performed in Biblioteca Medicea Laurenziana (September 2018)," *Analecta Papyrologica*, 2019, XXXI:301–316.

hand, it seems that the knowledge that iron sulphate (green vitriol) rather than copper sulphate (blue vitriol) is needed for production of black inks is far from universal in the 10th century CE.[23]

We hope that routine studies of the inks in a statistically relevant number of manuscripts originating over a larger period of time will reveal a coherent picture of the development and trade in the black writing inks. Moreover, we are confident that an extensive ink database would serve as a non-invasive dating and localizing tool for a large number of manuscripts of unknown provenance. Currently, no such a tool exists. In many cases, for which dating is crucial, we are limited to the invasive radioactive carbon method that dates organic writing surfaces rather than inks.

4 Conclusions

The great variety of writing inks appear in the centuries around the turn of the Christian era. The first step of our protocol developed for an easy ink recognition is pertinent for differentiating between pure-type inks, such as carbon, plant and iron-gall inks. Yet, the presence of metals in carbon-based inks can be discovered only if the second step of the protocol, i.e. the XRF analysis, is carried out. We therefore recommend including this step in the study of inks, also when a reflectographic test reveals the sooty nature of an ink. In the specific case that XRF analysis detects the presence of iron, we recommended using the third step of the protocol to differentiate between mere metallic iron and a proper mixture of soot and iron-gall ink.

23 See Claudia Colini's chapter in this volume (Chapter 7).

CHAPTER 5

'Alchemical' Inks in the Syriac Tradition

Matteo Martelli

Abstract

This paper explores recipes for ink making preserved in three Syriac alchemical manuscripts. First, I shall provide an analytical description of the scanty material transmitted in two codices kept at the British Library (Egerton 709 and Oriental 1593); then, particular attention will be devoted to a treatise that opens the collection of alchemical writings in the Cambridge MS Mm. 6.29 (15th century AD). This treatise includes several recipes on the making of inks that reveal evident similarities both with the instructions preserved in the Graeco-Egyptian tradition (especially in the so-called Leiden Papyrus) and with early medieval technical handbooks. A selection of Syriac recipes is edited here for the first time and translated and commented on in order to better understand the mechanisms that regulated the transmission of this technical material in Christian Near-Eastern communities.

Keywords

gold inks – Syriac alchemy – *Mappae clavicula*

1 The Syriac Alchemical Collections and the Making of Inks

In terms of the Syriac language,* alchemy is primarily known through the writings preserved in three manuscripts that date between the fifteenth and the sixteenth centuries: (1) British Library, MS Egerton 709 (16th century); (2) British Library, MS Oriental 1593 (15th–16th century); (3) Cambridge University

* This publication is part of the research project *Alchemy in the Making: From Ancient Babylonia via Graeco-Roman Egypt into the Byzantine, Syriac, and Arabic Traditions*, acronym *AlchemEast*. The *AlchemEast* project has received funding from the European Research Council (ERC) under the European Union's Horizon 2020 research and innovation programme (G.A. 724914).

DOI:10.1163/9789004444805_007

Library, MS Mm. 6.29 (15th century).[1] These collections, which contain recipes describing a wide array of technical procedures (e.g. preparations of dyeing substances and various chemicals; metallurgical processes; purple dyeing; imitation of gemstones, etc.) also include instructions on how to prepare inks of various colours. An interest in this area of expertise is not necessarily an unusual feature of the Syriac alchemical tradition, since recipes on the making of inks are included in many pre-modern collections of technical recipes related to alchemy. As we shall see, both the Graeco-Egyptian and the Latin traditions preserve recipe books that feature various ink formulas, in particular for metallic inks that allow for writing with gold and silver letters without actually using the precious metals (or by using a very small amount of them).

As Robert Halleux pointed out,[2] metallic inks were often produced by following the same procedures that were used to prepare the gold and silver paints that ancient alchemists used in their attempts to dye base metals and thus transmute them into precious metals. It is clear, therefore, that ink making was part of the wide umbrella of technologies that might be encountered in an ancient collection of writings dealing with alchemy or *chymeia* in Greek, a discipline that Byzantine sources have defined as "the preparation of gold and silver" (ἡ τοῦ ἀργύρου καὶ χρυσοῦ κατασκευή; *Suda* χ 280 Adler). A similar definition of alchemy is reported in the lexicon of the tenth-century Syriac polymath Bar Bahlūl, who described the art called *kimiyo* (ܟܝܡܝܐ, Syriac transliteration of the Greek χυμεία) as "the work of the art of gold and silver" (ܦܘܠܚܢܐ ܕܐܘܡܢܘܬܐ ܕܕܗܒܐ ܘܣܐܡܐ).[3] Indeed, like the Syriac alchemical collections that are the focus of this paper, various Late Byzantine alchemical compendia include recipes on the making of metallic inks. For instance, the Byzantine MS *Parisinus* gr. 2327 (15th century), one of the richest extant anthologies of late antique and Byzantine alchemical treatises, also includes a recipe book entitled "With the help of God, explanation of the most noble and illustrious art of goldsmiths."[4] This collection features three recipes for chrysography (i.e. writing with gold letters), which explain how to grind various metals (e.g. a bronze leaf that

1 See Rubens Duval and Marcelin Berthelot, *La chimie au Moyen Âge*, vol. 2. *L'alchimie syriaque* (Paris: Imprimerie Nationale, 1893), which includes a full edition and French translation of the collection preserved by the two British Library manuscripts along with a partial French translation of the texts transmitted in the Cambridge manuscript. See also Matteo Martelli, "L'alchimie en syriaque et l'œuvre de Zosime," in *Les sciences en syriaque*, edited by Émilie Villey (Paris: Geuthner, 2014), pp. 191–214.

2 Robert Halleux, *Papyrus de Leyde, Papyrus de Stockholm, Recettes* (Paris: Les Belles Lettres, 1981), p. 42.

3 Rubens Duval, *Lexicon Syriacum auctore Hassano bar Bahlule*, 3 vols. (Paris: Leroux, 1888–1902), vol. 1, p. 901.

4 Text edited in Marcelin Berthelot and Charles-Émile Ruelle, *Collection des anciens alchimistes grecs*, 3 vols. (Paris: Georges Steinheil, 1887–1888), vol. 2, pp. 321–337.

looks like gold or a gold-silver alloy) and mix them with different kinds of gluey substances, such as honey, egg white, and gum arabic.[5] Moreover, the same Byzantine manuscript also preserves formulas for different glues, which could be used in the preparation of metallic inks. For instance, a recipe preserved at the beginning of the manuscript describes the making of a cheese glue (fol. 7r; περὶ τοῦ ποιῆσαι τυρόκολλαν), which involves roasting old cheese, boiling it in water, and then mixing it with quicklime, in order to make a paste that can "glue whatever you want" (καὶ κόλλα εἴ τι δ' ἂν θέλῃς).[6]

2 The British Library Collection

The two Syriac manuscripts kept at the British Library preserve the same collection of ten alchemical books followed by various excerpts in *garšūnī*, which appear in the second part of the codices. This collection contains only scanty references to inks along with two short recipes in *garšūnī*, which describe the preparation of gold inks. Both recipes are included in the second part of the manuscripts, which features various technical passages on a variety of subjects, from the treatment of different metals and minerals to the classification of substances and instruments used by ancient alchemists. The first recipes in the *garšūnī* section describe various metallurgical techniques, among which is a recipe that explains how to liquefy a gold leaf in gum arabic (MSS Egerton 709, fol. 57r6–10; Oriental 1593, fol. 30r19–23):[7]

ܚܠ ܐܠܕܗܒ. ܐܢ ܐܪܕܬ ܬܚܠ ܐܘ ܬܠܝܢܗ. ܘܚܠ ܐܠܨܡܓ ܐܠܥܪܒܝ ܚܠܐ ܒܡܐ[8]
ܡܠܚ ܐܠܥܡܠ. ܘܐܠܩܝ ܒܗ ܐܠܘܪܩ. ܘܚܪܟܗ ܒܟܦܟ ܝܕܘܒ ܐܠܕܗܒ. ܘܣܩܗ[9]
ܒܐܠܨܡܓ ܚܠܐ ܡܚܠܐ. ܘܐܟܬܒ ܒܗ.

5 Berthelot and Ruelle, *Alchimistes grecs* (cit. note 4), vol. 2, pp. 327 (rec. 19–20), 334–335 (rec. 49). On these recipes, see also Peter Schreiner and Doris Oltrogge, *Byzantinische Tinten-, Tuschen- und Farbrezepte* (Wien: Verlag der Österreichischen Akademie der Wissenschaften, 2001), pp. 50–51, 58–59, 67.

6 Berthelot and Ruelle, *Alchimistes grecs* (cit. note 4), vol. 2, p. 380; Schreiner and Oltrogge, *Byzantinische Tinten* (cit. note 5), pp. 77–78. See also Maria Leontsini and Gerasimos Marianos, "From Culinary to Alchemical Recipes. Various Uses of Milk and Cheese in Byzantium," in *Latte e Latticini. Aspetti della produzione e del consumo nelle società mediterranee dell'Antichità e del Medioevo*, edited by Ilias Anagnostakis and Antonella Pellettieri (Lagonero: Grafica Zaccara, 2016), pp. 205–222 (pp. 216–217).

7 Syriac text edited in Berthelot and Duval, *Chimie* (cit. note 1), pp. 61–62; French translation on p. 142.

8 ܒܡܝ MS Oriental 1593.

9 MS Egerton 709 transmits the reading ܘܡܣܩܗ, corrected in ܘܣܩܗ ("it is watered with").

> Solution of gold (Ar. *ḥall al-ḏahab*). Take a porcelain bowl or a stone cutting board.[10] Gum arabic is scratched until it becomes like honey. Rub the bowl with it. A leaf of gold is flattened on it and it is crushed with the gum until it gets thinner.[11] Write with it.

This recipe has been copied on fol. 57r of MS Egerton 709 (see Fig. 5.1), which also includes marginal notes written by a later hand. In the upper margin an anonymous copyist has specified (presumably with reference to the above-mentioned recipe): عمل ماء ذهب للكتابة, "The making of gold 'water' for writing." Moreover, in the right margin, a second recipe has been added on the same subject. The text, difficult to interpret in various passages, reads:

> ܝܘܟܕ ܕܗܒ ܘܢܐܘܢܓ ܘܝܛܪܩ ܘܝܪܩܩ ܓܕܐ ܘܝܥܘܓ ܟܥܘܐܬ ܘܝܟܒܣ. ܬܡ ܝܓܥܠ ܦܝ
> ܒܘܛ ܩܪܝܒ ܐܠܢܐܪ ܘܝܓܥܠ ܦܘܩܗ ܐܠܙ. ܬܡ ܝܓܥܠ ܐܠܓܡܥܐ ܘܢܨܦ ܦܝ ܩܦܦܗ
> ܣܘܪܗ ܘܝܕܪܙ ܘܝܓܝ ܣܠܗ ܗܘܟܝ ܐܠܐ ܢܥܡ ܐܠܝܘܡ ܟܐܪܙ ܡܥܟܘܣ. ܦܝܓܝ ܡܐ
> ܐܠܥܡܠ ܘܝܟܬܒ.

> Gold is taken [...][12], hammered, made really thin, curved [...] and pressed; then it is put in a crucible[13] near to the fire and mercury (? or mastic) is placed over it. Then a handful (?)[14] and a half (of this compound?) is

10 The term ܨܠܐܝܗ corresponds to the Arabic صلاية, "a stone upon which one bruises, or powders perfume or some other thing," according to the definition in Edward William Lane, *An Arabic-English Lexicon*, 8 vol. (London: Williams & Norgate, 1863–1893), vol. 4, p. 1722; see also Rubens Duval, "Notes de lexicographie syriaque et arabe," *Journal asiatique*, 1893, 9th series, 2: 290–358 (p. 348).

11 I interpreted the verb ܢܚܠ as equivalent to the Arabic نحل, 'to be emaciated, grow thin.' Likewise, Berthelot and Duval, *Chimie* (cit. note 1), p. 142, translated: "jusqu'à ce que la feuille se délaye." We must note that the Syriac verb ܢܚܠ (like the Arabic نخل) means 'to sift, to sieve out'; this verb occurs in a recipe on the making of a black ink preserved in MSS Mingana syr. 77 (fol. 10v) and Mingana syr. 324 (fol. 75v–76r): edition and translation in Jimmy Daccache and Alain Desreumaux, "Les textes des recettes d'encres en syriaque et *garshuni*," in *Manuscripta syriaca. Des sources de première main*, edited by Françoise Briquel Chatonnet and Muriel Debié (Paris: Geuthner, 2015), pp. 195–246 (p. 212, rec. 34).

12 The term ܢܐܘܢܓ has been interpreted by Duval (who read ܢܐܪܢܓ in the MS) as a misspelled transcription of the Arabic نارنج, 'bitter orange.' He proposed the translation: "On prendra de l'or de teinte orange." See Berthelot and Duval, *Chimie* (cit. note 1), pp. 104 and 201. The form ܢܐܪܢܓܝ (= Ar. نارنجي) is attested in a recipe on the making of an orange ink (ܚܒܪ ܢܐܪܢܓܝ) recently edited and translated by Daccache and Desreumaux, *Recettes d'encres* (cit. note 11), p. 217.

13 On the term ܒܘܛ, 'crucible,' see Duval, *Lexicographie* (cit. note 10), p. 342.

14 I tentatively read the term ܐܠܓܡܥܐ as a misspelled transcription of الجماعة. In another *garšūnī* recipe edited by Berthelot and Duval, *Chimie* (cit. note 1), p. 99, l. 10, we read

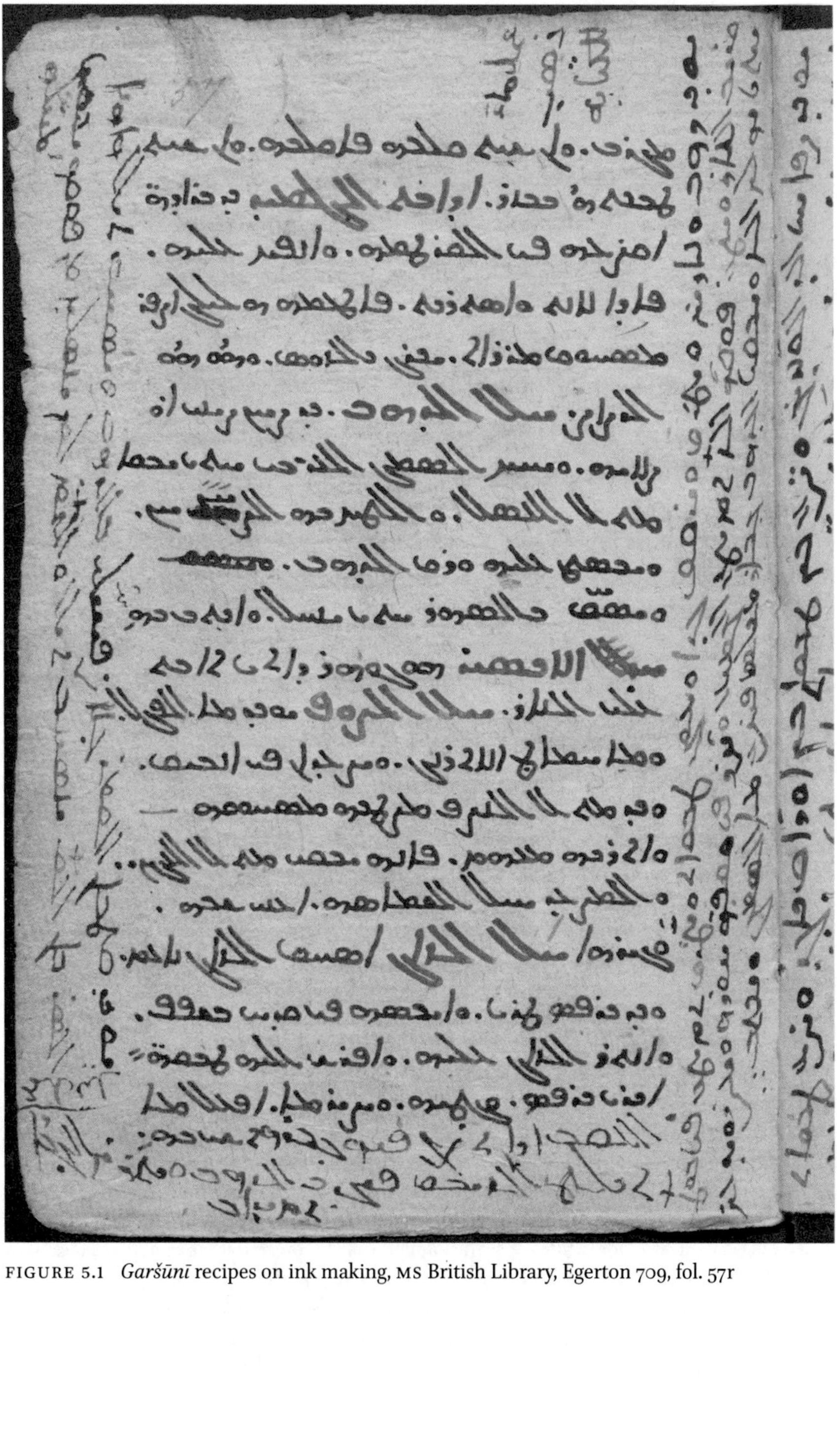

FIGURE 5.1 *Garšūnī* recipes on ink making, MS British Library, Egerton 709, fol. 57r

placed in […],[15] it is squeezed and crushed until the mercury (? or mastic) is destroyed and gold remains as a calcified dust; water of gum is then added, and you write.

At line 2, the term ܠܟ is uncertain: it might simply mean 'mastic,' as suggested by Duval,[16] or be interpreted as an abbreviated form for ܠܙܝܒܩ (الزئبق), 'mercury.' In fact, a sign — namely, a little cross — after this term refers to a second note written in the lower margin, which specifies that mercury was added in the treatment:

ܠܟܘܡܐ. ܐܘܐ ܬܢܨ ܒܗ ܥܦܪܗ ܘܬܚܠܛ ܠܙܝܒܩ ܡܬܓܪܓ ܠܕܗܒܐ ܘܡܣܓܐ ܬܘܒ ܡܫܬܪܐ.

What follows. If you keep[17] its dust (?) in it and mix mercury, the gold is attracted and increases, then it is dissolved.

The aim of this note is not entirely clear: the anonymous copyist seems to have recorded another section of the recipe, which explains how to dissolve gold by adding mercury. I have followed Berthelot and Duval's reading and interpreted the term ܥܦܪܗ as a variant spelling of ܥܦܐܪ (= Ar. عفار, 'dust'):[18] mercury, therefore, would be added to the gold, after its calcination. However, as already noted, in the manuscript this marginal annotation is linked to the term ܠܟ, perhaps to clarify that mercury was already to be added at this point of the

ܟܡܫܬܐ ܕܫܚܬܐ ܕܦܪܙܠܐ, translated as: "une poignée de scories de ferre" (p. 194). See Duval, *Lexicographie* (cit. note 10), p. 310.

15 This expression remains obscure to me (it might refer to a vessel or a tissue). Berthelot and Duval left it untranslated.

16 See Duval, *Lexicographie* (cit. note 10), pp. 305 and 319 (*s.v.* ܠܟ).

17 Berthelot and Duval, *Chimie* (cit. note 1), p. 201, n. 4, translated: "Si tu y mets de la poussière." They probably interpreted ܬܢܨ (= Ar. تنص) as a misreading of ܬܨܒ (= Ar. تصب). In this case, the copyist would have relied on a recipe written in Arabic script, where he read ن instead of ب.

18 Another possible interpretation would be to read ܥܒܝܪܐ (= Ar. غبيراء), 'rowan, mountain ash,' a plant that appears among the ingredients of a gold ink in a recipe preserved by MSS Mingana Syr. 77 (fol. 13r), Mingana syr. 314 (fol. 78r), and Berlin Sachau 107 (fol. 48v) under the title ܚܒܪܐ ܡܢ ܛܪܦܐ ܕܥܒܝܪܐ, "ink made from rowan leaves". Rowan leaves are mixed with gum and white lead; edition and translation in Daccache and Desreumaux, *Recettes d'encres* (cit. note 11), p. 215. See also Philippe Boutrolle and Jimmy Daccache, "Lexique commenté: les végétaux, les animaux et les minéraux des recettes d'encres en syriaque et en *garshuni*," in *Manuscripta syriaca. Des sources de première main*, edited by Françoise Briquel Chatonnet and Muriel Debié (Paris: Geuthner, 2015), pp. 247–270 (pp. 258–259).

process. As we shall see in other recipes for the making of gold inks, gold leaves are often processed with mercury.[19]

The two short texts discussed above represent the main sources on ink making in the collection disclosed by the British Library manuscripts, both included in a *garšūnī* section that collates many procedures for 'liquifying' either metals or minerals. Some metals (e.g. tin or lead) were simply melted, while other minerals like sandarac and vitriol underwent more complex treatments. The production of inks was somehow assimilated to these techniques: gold was not simply melted but crushed to a 'powder' or 'dust' and then mixed with a gluey liquid. On the other hand, if we return to the first part of the British Library collection, the ten Syriac alchemical books do not include any recipe dealing with the production of inks. Black inks, however, do appear in these books, as ingredients used in alchemical procedures. The first explicit reference to writing practices is encapsulated in the list of alchemical signs and abbreviations that begins the ten alchemical books. This list records either alchemical symbols — signs referring to metals, plants and minerals (e.g. electrum, vinegar, lime, copper, realgar, alum, lac-dye, etc.) — or transcriptions of Greek terms, which are written in red ink and followed by an explanatory note in black (see MS Egerton 709, fol. 2r–5v and Oriental 1593, fol. 1r–3v). Although similar, this list reveals relevant variants within the two British Library manuscripts. Only MS Oriental 1593 features the following entry (fol. 1r14):

ܡܥܢܝܢ. ܕܝܘܬܐ ܕܟܬܘ̈ܒܐ.

M'NYN (*lege MLNYN*): ink of the writers.

The lemma *M'NYN* seems to be a misspelled transcription of the Greek term μελάνιον,[20] the diminutive of μέλαν (lit. 'black,' 'ink'), which is transcribed as

19 The margin of MS Egerton 709, fol. 56v, preserves a few lines of another recipe that describes how to hammer gold leaves: Berhelot and Duval, *Chimie* (cit. note 1), p. 103. Next to these lines, a note in Arabic reads: عمل ورق ذهب للنقش, "The working of gold leaves for the painting."

20 The term is attested in a recipe on the preparation of a 'magical' ink (τοῦ μελανίου ἡ σκευή) included in *PGM* I 253 (Karl Preisendanz, *Papyri Graecae magicae. Die griechische Zauberpapyri*, Stuttgart: Teubner, 1973, vol. 1, p. 14): see Miriam Blanco's contribution in this volume (Chapter 2). A substance called μελάνιον is also mentioned in the *Libri medicinales* (*Medical books*) of the 6th-century Byzantine physician Aetius of Amida: in book XVI 146,12, μελάνιον appears among the ingredients of a scented drug to be used in fumigation (Skevos Zervos, *Gynaekologie des Aetios sive sermo sextus decimus et ultimus zum erstenmale aus Handschriften veröffentlicht*, Leipzig: Verlag von A. Fock, 1901, p. 170 = Roberto Romano, "Aezio Amideno libro XVI," in *Medici bizantini*, edited by Antonio

MHL'WN (ܡܗܠܐܘܢ) or *MHL'N* (ܡܗܠܐܢ) in other Syriac texts:[21] rather than *M'NYN*, here we should probably read *MLNYN* (ܡܠܢܝܢ), as already suggested by Duval and Brockelmann.[22] This correction is confirmed by an entry included in a Galenic section preserved in the Cambridge University Library alchemical manuscript.[23] The manuscript, in fact, includes a summarized Syriac translation of Galen's treatise *On Simple Drugs*, books 9–11. The translation takes the form of a lexicon attributed to the Greek alchemist Zosimus of Panopolis. The Greek names of various minerals and animal products originally described by Galen are simply transliterated in Syriac script and briefly explained. In particular, in book 9 of *On Simple Drugs*, Galen devoted two entries to the medical uses of μελαντηρία, 'shoemakers' black' (book 9, chap. 3.19 = 12.226, 4–6 Kühn) and μέλαν, 'ink' (book 9, chap. 3.20 = 12.226, 7–10 Kühn).[24] These two substances appear to have been combined in a single entry in the Syriac summary, which reads (MS Mm. 6.29, fol. 121r24–25):

ܡܐܠܐܢܛܪܝܐ ܕܐܝܬܘܗܝ ܡܠܢܝܢ [...] ܕܝܘܬܐ ܕܟܬܘ̈ܒܐ.

> *M'L'NṬRY'* (= Gr. μελαντηρία), that is *MLNYN* (= Gr. μελάνιον), [...][25] ink of the writers.

The term *MLNYN*, based on the Greek diminutive form μελάνιον, might have been used here to 'translate' the Greek μέλαν: indeed, as recently argued by

Garzya, Torino: UTET, 2006, p. 548). See also Erich Trapp, *Lexikon zur byzantinischen Gräzität besonders des 9.–12. Jahrhunderts* (Wien: VÖAW, 2001–2017), vol. 2, *s.v.* μέλανιν.

21 Robert Payne Smith, *Thesaurus Syriacus* (Oxford: Clarendon Press, 1897–1901; hereafter *ThSyr*), p. 2025. For instance, the 10th-century lexicographer Bar Bahlūl provides the following explanation: ܡܗܠܐܢ. ܕܝܘܬܐ مداد الكتّاب, "*MHL'N* (= Gr. μέλαν), ink (Syriac *dyuto*), ink of the writers (Arabic *midād al-kuttāb*)," see Duval, *Lexicon Syriacum* (cit. note 3), vol. 2, p. 1022, l. 6.

22 Berthelot and Duval, *Chimie* (cit. note 1), p. 11, n. 1, and Carl Brockelmann, *Lexicon Syriacum*, 2nd ed. (Halis Saxonum: Sumptibus M. Niemeyer, 1928), p. 302. See Michael Sokoloff, *A Syriac Lexicon. A Translation from the Latin, Correction, Expansion and Update of C. Brockelmann'* Lexicon Syriacum (Winona Lake, IN–Piscataway, NJ: Eisenbrauns–Gorgias Press, 2009), p. 774.

23 MS Mm. 6.29, fol. 120v–121v + 148 + 122r–129v. See Matteo Martelli, "Medicina e alchimia. 'Estratti galenici' nel *Corpus* degli scritti alchemici siriaci di Zosimo," *Galenos*, 2010, 4: 207–228; Martelli, *L'alchimie syriaque* (cit. note 1), pp. 208–211.

24 Uncritical edition by Karl G. Kühn, *Claudii Galeni opera omnia*, vol. 12 (Leipzig: Knobloch, 1826).

25 The manuscript is damaged here and a word has been erased by humidity; Berthelot and Duval, *Chimie* (cit. note 2), p. 298, paraphrased the whole entry as follows: "μελαντερία [*sic*!] ou μέλαν, c'est l'encre des écrivains."

Aaron Butts, in many cases "the diminutive serves as an input form for a number of Greek loanwords in Syriac."[26] Moreover, both in the opening list of the MS Oriental 1593 and in the Galenic section ascribed to Zosimus, the term *MLNYN* is explained with *dyuto* (ܕܝܘܬܐ), a standard Syriac term for 'ink,'[27] which is used with reference to inks of various colours in many recipes recently edited and commented on by Jimmy Daccache and Alain Desreumaux.[28] Our alchemical passages further specify that the ink was used by 'writers' (ܣܦܪ̈ܐ), presumably either private or professional copyists working in Christian *scriptoria*.

The inclusion of terms referring to inks and writing practices in a list of alchemical signs and Greek loanwords might astonish the reader. This apparently unexpected mention, however, does not seem out of the ordinary, especially when the Syriac list is compared with the Byzantine tradition. Byzantine alchemical manuscripts, indeed, also feature lists of signs that reveal many similarities with the opening sections of the two British Library codices. In particular, the Byzantine manuscripts *Parisinus gr.* 2327 (1478 AD) and *Laurentianus Plut.* 86, 16 (1492 AD) feature long lists of alchemical signs that include a reference to the Greek term μέλαν and its abbreviation.[29] Moreover, a Byzantine lexicon of alchemical terms entitled *Lexicon on the Making of Gold*, organized alphabetically, features the following entry under the letter μ: "Indigo is prepared with woad and chrysolite" (μέλαν Ἰνδικόν ἀπὸ ἰσάτιος γίνεται καὶ χρυσολίθου).[30] The expression clearly refers to a blue-dark colour; indeed, indigo (μέλαν Ἰνδικόν) also appears among the ingredients used to produce an artificial hyacinth in a Byzantine recipe book on the making of gemstones.[31] On the other hand, the black ink of writers (μέλαν γραφικόν) is explicitly mentioned by the Graeco-Egyptian alchemist Pelagius to describe

---

26 Aaron Michael Butts, *Language Change in the Wake of Empire. Syriac in Its Greco-Roman Context* (Winona Lake, IN: Eisenbrauns, 2016) p. 101.

27 See, for instance, Bar Bahlūl explanation: ܕܝܘܬܐ. المداد الحبر, "Ink (Syr. *dyuto*), ink (Ar. *al-midād, al-ḥibr*)," see Duval, *Lexicon Syriacum* (cit. note 3), vol. 1, p. 562, l. 6.

28 Daccache and Desreumaux, *Recettes d'encres* (cit. note 11). See also Alain Desreumaux, "Des couleurs et des encres dans les manuscrits syriaques," in *Manuscripta syriaca. Des sources de première main*, edited by Françoise Briquel Chatonnet and Muriel Debié (Paris: Geuthner, 2015), pp. 161–192 (p. 181).

29 For the list in the MS *Laurentianus Plut.* 86,16, see Zuretti, *Catalogue des manuscrits alchimiques grecs*, vol. 7. *Alchemistica signa* (Bruxelles: Union Académique Internationale, 1932), p. 16, l. 863; for the MS *Parisinus* gr. 2327, see Berthelot and Ruelle, *Alchimistes grecs* (cit. note 4), vol. 1, p. 114, l. 4.

30 Berthelot and Ruelle, *Alchimistes grecs* (cit. note 4), vol. 2, p. 11, l. 6. On this lexicon, see also Matteo Martelli and Stefano Valente, "Per una nuova edizione di un lessico alchemico bizantino," *Eikasmos*, 2013, 24:275–296.

31 Berthelot and Ruelle, *Alchimistes grecs* (cit. note 4), vol. 2, p. 351, l. 29.

a black compound produced by treating chrysolite and *magnēsia*.[32] Likewise, in the collection of Syriac alchemical books preserved in the British Library manuscripts, ink (*dyuto*) is introduced among the ingredients used for specific preparations. For instance, purified 'Indian ink' (ܕܝܘܬܐ ܗܢܕܘܝܬܐ, namely 'Indigo'), 'flower of copper' (ܗܒܒܐ ܕܐܦܪܘܪ), verdigris (ܚܡܣܝܪܐ) and juice of leeks (ܚܘܪܐ ܕܟܪ̈ܬܐ) represent the main ingredients of a red alchemical 'water' whose preparation is described in the ninth book of the collection.[33]

3 The Cambridge Alchemical Manuscript: An Overview of Ink Recipes

The scattered references to inks discussed so far are relics of a broader interest in ink making technologies (especially the making of gold and silver inks) that emerges more clearly in the collection of alchemical writings preserved by the Syriac MS Mm. 6.29. The first folia of the manuscript have been lost. In its current state, the collection opens with a series of recipes on metallurgical procedures that are difficult to read: fol. 1r, in fact, has been heavily damaged by humidity. The section is closed with the explicit: ܫܠܡ ܡܐܡܪܐ ܩܕܡܝܐ ܕܥܠ ܨܒ̈ܥܐ , "End of the first treatise on colours."[34] Then, a second book begins (fol. 1v3), introduced by the title: ܡܐܡܪܐ ܕܬܪܝܢ. ܗܦ[...] ܐܬܘܬܐ ܒܝܬ ܕܡܡܠܠܐ ܥܠ ܟܬ̈ܒܬܐ ܕܟܠ ܓܢܣ ܘܥܠ ܛܘܫ̈ܐ ܕܢܟܬܘܒ ܐܝܟ ܒܕܗܒܐ, "Second treatise, *hp*(…), letter *bēt* that deals with letters of any kind and paints (lit. 'coating') to write with gold." The text — as we shall see, a collection of recipes — seamlessly continues until fol. 20v, where its end is marked by the *explicit*: ܫܠܡ ܡܐܡܪܐ ܩܕܡܝܐ, "End of the first treatise." The inconsistency of this ending with the title is evident; indeed, we would have expected a reference to either book 2 or to the second letter of the Syriac alphabet (i.e. *bēt*). Hence, Berthelot and Duval supposed that this section actually merges two originally separate books: (1) a first treatise on gold inks and paints, which runs until fol. 9r, where it concludes with a recipe on how to write on iron with black letters (fol. 8v21–9r8: ܠܟܬ̈ܒܬܐ ܐܘܟܡ̈ܬܐ ܥܠ ܐܪܙܢܐ); (2) an untitled book on various metallurgical procedures and dyeing techniques, which ends with the above-mentioned *explicit*.[35] Indeed, after this first

32 *Ibid.*, p. 255, l. 20: γίνεται δὲ πάνυ μέλαν ὡς τὸ γραφικὸν μέλαν, "it becomes completely black as the ink for writing." On Pelagius, see Jean Letrouit, "Chronologie des alchimistes grecs," in *Alchimie: Art, histoire et mythes*, edited by Didier Kahn and Sylvain Matton (Paris–Milano: S.É.H.A.–Arché, 1995), pp. 11–93 (pp. 46–47).

33 Berthelot and Duval, *Chimie* (cit. note 1), p. 49 (see l. 8 in particular).

34 *Ibid.*, p. 203.

35 *Ibid.*, pp. 209–210.

treatise, the manuscript continues with other treatises progressively marked by Syriac letters in alphabetical order, from the second (*bēt*) to the eleventh letter (*kop*), which appear to be the Syriac translation of original Greek texts by the Graeco-Egyptian alchemist Zosimus of Panopolis (3rd–4th century AD).[36]

The possible relationship of the first two books with the other treatises by Zosimus awaits a proper examination and, hopefully, the question will be better assessed after the publication of a complete edition of Zosimus' Syriac books.[37] Here, it will suffice to note that these books include various references to the use of compounds or simple substances as inks. Book 9 (on mercury), for instance, contains a recipe on how to produce a "golden mercury" (Mm 6.29, fol. 58v19–20) by grinding nails of gold (ܙ̈ܘܙܐ ܕܕܗܒܐ) in a mortar and mixing it with water or liquid gum (? ܫܘܦܬܐ).[38] Thanks to this procedure, the author specifies (fol. 58v19–20), ܐܝܬ ܠܟ ܗܘ ܙܝܘܓ ܕܕܗܒܐ ܕܚܫܚ ܐܦ ܠܟܬܒ̈ܐ [...], "You will have the mercury of gold that is useful also for the books (i.e. for writing on books)." In book 11 (on iron), a recipe on the treatment of iron is introduced by the title: ܐܣܘܬܐ ܕܐܦܪܙܠ ܕܐܝܟܢܐ ܬܟܬܘܒ ܒܗ ܥܠ ܙܓܘܓܝܬܐ, "Preparation of iron with which you will write on glass" (Mm. 6.29, fol. 77v24–28).[39] Finally, a section of book 7 explains a gilding technique that could be applied to both metallic leaves and parchment (Mm. 6.29, fol. 48v13–21). The recipe describes how to prepare a golden paint without actually adding the precious metal: a wide variety of ingredients — such as lime (ܓܦܣܘܣ = Gr. γύψος), fish glue (ܕܒܩܐ ܕܢܘܢܐ), *MLYSYN* (ܡܠܝܣܝܢ = Gr. Μιλήσιον),[40] ochre (ܐܘܟܪܐ = Gr. ὤχρα) and minion (ܡܝܢܪܐ) — are mixed together in different steps and applied to metallic leaves. The same product could also be used to write on parchment (ܒܡܫܟܐ ܕܟܬܒ̈ܐ).[41]

The wide range of substances handled by ancient alchemists included 'chemicals' that could serve multiple applications. Similar paints were used both to

36 See also Martelli, *Alchimie en syriaque* (cit. note 1), pp. 199–209.

37 The edition is in progress as part of the ERC project *AlchemEast*.

38 Berthelot and Duval, *Chimie* (cit. note 1), p. 245, translate: "avec de l'eau ou de la gomme liquide." See also Duval, *Lexicographie* (cit. note 10), p. 366. The usual meaning of ܫܘܦܬܐ is, however, 'filings.'

39 Berthelot and Duval, *Chimie* (cit. note 1), p. 257.

40 According to ancient Greek medical literature, the adj. Μιλήσιος 'Milesian' was used to qualify a specific kind of ἀλκυόνιον (perhaps a type of coral or sponge). In his book *On Materia Medica*, for instance, Dioscorides writes (V 118): "You must know that there are five kinds of *alkyonia*. One kind is thick, sponge-like in structure and heavy; [...] the third is vermicular and of purplish colour; some call it Milesian (τὸ δὲ τρίτον σκωληκοειδὲς ὑπάρχει τῷ τύπῳ, καὶ τῇ χρόᾳ ἐμπόρφυρον, ὅ τινες Μιλήσιον καλοῦσι)." Translation by Lily Y. Beck, *Pedanius Dioscorides of Anazarbus, De Materia Medica*, 2nd edition (Hildesheim–Zürich–New York: Olms–Weidmann, 2011), p. 392.

41 Berthelot and Duval, *Chimie* (cit. note 1), p. 237.

treat different objects and to write on metals as well as on books. The variety of the supports to which paints and inks could be applied clearly emerges in the above-mentioned book entitled "Second treatise, *hp*(…), letter *bēt* that deals with letters of any kind and paints to write with gold." The book devotes special attention to the making of metallic inks that could be used to paint or write on metallic vessels, glass, marble, wood, paper, and parchment. On the other hand, a broader scope that also includes black inks is detectable in another treatise preserved in MS Mm. 6.29, in a second section transmitted under the name of the Greek philosopher Democritus (fol. 90v1–116v5). Democritus' reputation as an alchemist can be traced back to the Graeco-Roman period, and the Byzantine tradition preserves excerpts of four alchemical books under his name, dating to the first century AD.[42] The Syriac translation of relevant sections of these books is transmitted in the Cambridge manuscript (fol. 90v1–98r2),[43] which also includes three further treatises attributed to the philosopher (fol. 98r14–116v5).[44] The first of these treatises contains various recipes on the making of ink, including the following formula (Fol. 98r16–19):

> ܣܒ ܠܟ ܐܩܪܐ ܩܝܩܝܕܘܢ ܘܕܘܩܘܩܝ ܛܒ. ܘܐܪܡܐ ܥܠܘܗܝ ܡܝܐ ܘܟܬܘܒ ܙܘܡܘܣ ܒܠܗܘܢ
> ܥܠ ܟܪܛܝܣܐ ܘܐܪܦܐ ܠܗ ܕܢܒܫ ܘܣܒ ܙܘܡܐ ܕܡܘܣܘܣ ܘܐܪܒܝܥ ܒܗ ܠܟܪܛܝܣܐ ܘܩܪܝ.
>
> Take gallnuts used for inks (= Gr. κηκίδιον) and pound them well; pour water over it, write on paper with this liquid (= Gr. ζωμός, lit. 'sauce, wash') and let it get dry; take a liquid of *misy* (an iron-copper ore), moisten the paper with it, and read.

The recipe describes a common technology in Antiquity, namely the making of black inks by mixing oak galls with iron ores. Various formulas of iron-gall inks are preserved in Syriac manuscripts from the 10th to 19th centuries.[45] The recipe edited here, however, testifies to a specific iteration of this procedure, designed to write invisible letters that become visible when moistened with liquids rich in iron. A similar trick is already attested by sources dating back

42 See Matteo Martelli, *The Four Books of Pseudo-Democritus* (Leeds: Maney Publishing, 2013).

43 Syriac text edited, translated, and commented on in Martelli, *The Four Books* (cit. note 42), pp. 152–187, 251–263.

44 Partial French translation in Berthelot and Duval, *Chimie* (cit. note 1), pp. 275–293.

45 See Daccache and Desreumaux, *Recettes d'encres* (cit. note 11); Desreumaux, "Des couleurs et des encres" (cit. note 28), pp. 181–182. For a similar preparation, see recipes Nos. 16 and 39 in Lucia Raggetti's contribution in this volume (Chapter 8).

to the Hellenistic period. In his *Compendium on Mechanics* (IV 77),[46] the 3rd-century BCE writer Philo of Byzantium mentions a special ink made of gall nuts dissolved in water, which enabled the writing of letters that became invisible as the mixture dried; however, the letter becomes legible again once it is washed with a sponge soaked in a solution of vitriol.[47]

4 The Book on Inks in the Cambridge Manuscript

As already mentioned, the first section of the MS Mm. 6.29 preserves a treatise marked by the letter *bēt*, which collates more than 40 recipes on the making of inks.[48] The compiler's interest appears to be primarily focused on golden inks, which often include a quantity of gold. In some cases, however, cheaper ingredients — either metallic alloys or yellow plants and minerals — are used to prepare products that do not require the use of the precious metal. Methods for writing black letters on metallic surfaces are also described, along with those for the preparation of silver inks. Many of the recipes included in this book are based on earlier Greek texts that have been preserved in their original language in only a few cases. Indeed, the description of some procedures in the Syriac book matches the technical information conveyed by the Leiden Papyrus X (3rd–4th century AD), a Greek collection of (al)chemical recipes that also features various formulas on the preparation of gold and silver inks.[49] However, the most important alchemical source that overlaps significantly with the Syriac treatise is a Latin recipe book usually referred to as the *Mappae clavicula*. This has been identified as an early Medieval translation of a lost Late Antique alchemical treatise originally written in Greek.[50]

---

46 Edition in Hermann Diels and Erwin Adelbert Schramm, *Exzerpte aus Philons Mechanik B. VII und VIII (vulgo fünftes Buch). Abhandlungen der preußischen Akademie der Wissenschaften, Philosophisch-historische Klasse Nr. 12.* (Berlin: Reimer, 1920), p. 79.

47 Robert J. Forbes. *Studies in Ancient Technology*, 9 vols (Leiden: Brill, 1966–1993), vol. 3, pp. 236–239; Thomas Christiansen, "Manufacture of Black Ink in the Ancient Mediterranean," *The Bulletin of the American Society of Papyrologists*, 2017, 54:167–195 (pp. 188–190).

48 Partial French translation in Berthelot and Duval, *Chimie* (cit. note 1), pp. 203–209.

49 See the recent edition and French translation by Halleux, *Papyrus* (cit. note 2), pp. 84–109 (see pp. 42–43 for an introduction to the recipes for ink making). A full English translation of the papyrus (although based on an earlier and less reliable edition) is available in Earle Radcliffe Caley, "The Leiden Papyrus X. An English Translation with Brief Notes," *Journal of Chemical Education*, 1926, 10:1149–1166.

50 For a recent edition and Italian translation of the *Mappae clavicula*, see Sandro Baroni, Giuseppe Pizzigoni, and Paola Travaglio, *Mappae clavicula. Alle origini dell'alchimia in*

Scholars have already pointed to various parallels between these Greek, Latin, and Syriac recipe books.[51] However, the lack of a proper edition of the Syriac treatises has, so far, ruled out a close examination of these similarities. A full investigation in this sense would certainly go beyond the scope of this paper. In what follows, I will focus rather on the first part of the Syriac book (recipes 1–16), by providing an analytical description of its contents and, when possible, by comparing the Syriac texts, edited here for the first time, with the Greek and Latin traditions.

The Syriac book opens with a series of recipes on the making of golden inks, which include the addition of the precious metal in various forms (nails, filings, leaves):

(R1) Fol. 1v5–16 (golden ink with the precious metal)
Inc.: ܣܒ ܦܐܠܝܡ, "take a cup (= Gr. φιάλη)."

(R2) Fol. 1v17–22
Another recipe, difficult to read (the text is damaged); in the margin: ܣܝܪ (*sir*).[52]

(R3) Fol. 2r8–13 (golden ink with the precious metal)
Inc.: ܣܝܪ. ܣܒ ܙܪ̈ܐ ܕܕܗܒܐ, "*sir*. Take nails of gold."

(R4) Fol. 2r13–21 (golden ink with the precious metal)
Inc.: ܣܝܪ. ܣܒ ܫܘܦܐ ܕܕܗܒܐ, "*sir*. Take filings of gold."

(R5) Fol. 2r21–2v3 (golden ink with the precious metal)
ܣܝܪ. ܣܒ ܫܘܦܐ ܕܕܗܒܐ ܘܐܪܙܐ ܠܠܒܝܢ. ܘܫܚܘܩ ܒܚܠܐ ܚܡܥܐ ܕܡܥܠܝܐ. ܘܚܠܛܘܗܝ
ܥܡ ܚܠܐ. ܘܣܒ ܕܘܚܐ ܕܢܘܪܐ ܘܫܘܦ ܒܨܒܥܐ ܘܟܬܘܒ ܒܩܢܝܐ ܘܢܬܦܣ
ܒܨܒܥܐ ܘܚܠܛܘܗ.

Occidente (Saonara: Il prato, 2013). On the Greek origins of the Latin treatise, see Robert Halleux and Paul Meyvaert, "Les origines de la *Mappe clavicula*," *Archives d'histoire doctrinale et littéraire du Moyen Âge*, 1987, 54:7–58.

51 Robert Halleux, for instance, often refers to the *Mappae clavicula* and the Syriac collections of alchemical texts when commenting on single recipes in his edition of the Leiden papyrus: Halleux, *Papyrus* (cit. note 2), *passim*; on the other hand, a preliminary list of correspondences between recipes of the *Mappae clavicula* and the Syriac treatise has been drawn in Baroni et al., *Mappae clavicula* (cit. note 50), pp. 237–242.

52 The term — which seems to be an abbreviated form of ܣܝܪܐ (= Gr. σεῖρα, 'chain') — is often used in this collection to introduce recipes: see Duval, *Lexicographie* (cit. note 10), pp. 323–324.

Sir. Take filings of gold and put in a mortar. Grind them with vinegar until they become liquid. Then drip vinegar and add fish glue; grind with water, mix together, moisten with water, and write.

Here, gold is added to the preparation after being limed and reduced to filings. Indeed, filings were probably easier to grind in a mortar than gold leaves, which are very malleable and difficult to process. As we shall see, when gold leaves appear in ink recipes, they are often mixed with mercury to facilitate the grinding process.[53] In our recipe, vinegar seems to be used for the same purpose: the aim was to produce a gold powder in suspension in a liquid substance. Fish glue is mixed to make the ink stickier. A similar technique is already attested in the Leiden Papyrus, according to which a gold ink was produced by grinding a gold alloy with vinegar (rec. 44):[54]

Χρυσογραφία. Χρυσᾶ γράμματα γράφειν· κολλῇ χρυσοχοικῇ γράφε ὃ θέλεις σὺν ὄξει.

Chrysography. To write in letters of gold. Write what you desire with goldsmith's solder and vinegar.[55]

Rather than pure gold, a gold alloy is used here. Indeed, the Leiden Papyrus also includes two different formulas for gold solders, in which gold (in different proportions) is alloyed with copper and silver.[56] According to Alexander's experiments, who used a gold-copper alloy (2 parts of gold, 1 part of copper) with a little of silver,[57] "the ink — made by filing the metal, rising the powder in vinegar and mixing with gum — is reddish brown when applied to the parchment, though discrete particles of metal give a decided lustre to the surface."[58] The Greek text, however, fails to specify the addition of a glue, like gum or fish glue.

In some cases, litharge — an orange-yellow lead oxide — was also mixed with gold, as described in the next recipe of the Cambridge manuscript:

53 Shirley M. Alexander, "Medieval Recipes Describing the Use of Metals in Manuscripts," *Marsyas*, 1964–1965, 12:34–53 (p. 38).

54 Halleux, *Papyrus* (cit. note 2), p. 96.

55 Translation by Caley, *The Leiden Papyrus X* (cit. note 49), p. 1157.

56 See rec. 30 and 32 of the Leiden Papyrus: Halleux, *Papyrus* (cit. note 2), pp. 92–93.

57 This is the formula given by rec. 32 of the Leiden Papyrus.

58 See Alexander, *Medieval Recipes* (cit. note 53), p. 40.

(R6) Fol. 2v3–7 (golden ink with the precious metal)

ܣܪ. ܫܩܘܠ ܕܗܒܐ ܕܟܝܐ ܡܢܬܐ ܚܕܐ. ܡܪܬܟܐ [...] ܫܘܕ ܐܟܚܕܐ ܘܚܠܘܛ. ܘܒܬܪܟܢ ܫܚܘܩ
ܘܐܫܝܓ ܒܡܝܐ ܥܕܡܐ ܕܡܬܕܟܝܐ ܡܪܬܟܐ. ܘܐܪܡܐ ܡܝܐ ܩܘܡܝ ܘܟܬܘܒ.

Sir. Take pure gold, 1 part; litharge, [...] and pour together and mix, then grind and wash with water until litharge is purified; add water of gum and write.

The addition of lead (in the form of litharge, i.e. a lead oxide) could make the gold brittle, so that it was more easily filed to a fine powder. Moreover, the yellow colour of litharge made the substance more suitable for the preparation of golden inks.[59] A gold-lead alloy is used, for instance, in recipe 68 of the Leiden Papyrus, which prescribes melting the alloy before grinding it in a mortar of jasper and adding vinegar and soda.[60] As we shall see, a similar alloy is also used in recipe 13 of the Cambridge manuscript (see below).

After this first section, which only include formulas that require the addition of gold, two recipes **(R7–8)** record the complex formulas of golden inks (or paints), which involve a set of vegetal, mineral, and animal ingredients. Only the second recipe **(R8)** prescribes adding a metallic leaf, without specifying which metal it is made from:

(R7) Fol. 2v9–14 (golden ink without gold used to write on paper, glass and marble)

ܘܐܪܫܐ. ܡܪܪܬܐ ܕܟܫܐ. ܐܪܣܢܝܩܘܢ ܕܕܗܒܐ. ܙܪܩܐ ܕܡܫܝܚܐ.[61] ܓܘܐ ܕܩܠܦܐ ܕܪܘܡܢܐ
ܕܝܒܝܫܐ. ܐܪܥܐ ܕܣܡܘܣ. ܟܘܪܟܡܐ. ܠܟܠܗܘܢ ܣܒ ܚܕܐ ܚܕܐ ܡܢܬܐ. ܘܚܘܪܐ ܕܒܝܥܐ
ܚܡܫ ܘܥܒܕ ܓܒܝܠܐ. ܘܡܐ ܕܨܒܐ ܐܢܬ ܕܬܨܘܪ ܚܠܘܛ ܒܡܝܐ ܩܘܡܝ ܐܘ ܒܚܡܪܐ
ܕܙܪܩܢ.

W'RŠ' (a yellow dye plant; see Ar. ورس),[62] bile of a tortoise, golden orpiment (= Gr. ἀρσενικόν), scissile alum, inner part of the skin/peel of dried pomegranates, earth from Samos, saffron; for each of these (ingredients) take 1 part, and the white of 5 eggs, and reduce them to a paste. When you want to paint, mix it in water of gum or in reddish wine.

59 Rec. 34 of the Leiden Papyrus simply prescribes: "Another (recipe). Golden-coloured litharge, 1 part; alum, 2 parts." Caley, *The Leiden Papyrus* (cit. note 49), p. 1156; Halleux, *Papyrus* (cit. note 2), p. 93.

60 Alexander, *Medieval Recipes* (cit. note 53), p. 39. Text edited in Halleux, *Papyrus* (cit. note 2), p. 100.

61 The MS has ܙܡܫܚܐ.

62 Other spellings: ܘܪܫ and ܗܪܫ; see Duval, *Lexicographie* (cit. note 10), pp. 312–313; Sokoloff, *A Syriac Lexicon* (cit. note 22), p. 360.

(R8) Fol. 2v14–17 (golden ink)

ܣܝܪ. ܘܐܪܫܐ. ܐܪܣܢܝܩܘܢ ܕܫܦܝܪ ܘܡܕܟܝ. ܐܢܫ ܩܠܝܠ. ܝܘܪܓܢܐ. ܡܪܪܬܐ ܕܥܓܠܐ. ܩܘܡܝ
ܕܕܗܒܐ. ܡܢ ܟܠ ܚܕ ܡܢܗܘܢ ܚܕ ܚܕ ܡܢܝܐ. ܘܚܠܛ ܐܢܘܢ ܘܫܚܘܩ ܒܗܘܢ ܛܪܦܐ
ܘܟܬܘܒ.

Sir. W'RŠ' (a yellow dye plant), orpiment of good quality, that has been purified, a bit of verdigris, bile of a calf, golden gum; for each of these (ingredients), 1 part; mix these (ingredients), pound a (metallic?) leaf with them and write.

These two recipes exhibit a number of similarities with other texts preserved in the Leiden Papyrus, the *Mappae clavicula*, and in the above-mentioned book of Democritus that is handed down in the same Cambridge Manuscript and that includes recipes on ink making.[63] In fact, this book contains the following formula (MS Mm. 6.29 fol. 100r7–13):

ܐܚܪܢܐ. ܟܠܘܕܪܘܢ ܡܢܬܐ ܚܕܐ. ܐܪܣܢܝܩܘܢ [...]ܕܗܒܐ ܡܢܬܐ ܚܕܐ. ܡܪܪܬܐ ܕܥܓܠܐ
ܡܢܘܬܐ ܚܡܫ. ܗܠܝܢ ܕܝܢ ܟܠܗܘܢ ܕܪ̈ܟܡܘܢܐ ܥܣܪܝܢ. ܗܘܐ ܕܝܢ ܢܬܪܡܐ ܥܠܝܗܘܢ.
ܟܘܪܟܡܐ ܕܡܢ ܩܝܠܝܩܝܐ ܕܪ̈ܟܡܘܢܐ ܐܪܒܥ. ܡܢ ܗܢܐ ܟܬܘܒ ܥܠ ܐܝܢܐ ܕܨܒܐ ܐܢܬ.
ܒܡܐܢܐ ܘܒܟܪ̈ܛܝܣܐ ܘܐܦ ܒܟܐܦܐ ܘܒܩܝܣܐ.

Another (recipe). Celandine (= Gr. ἐλύδριον), 1 part; golden (?) orpiment (= Gr. ἀρσενικόν), 1 part; bile of a calf,[64] 5 parts. All these (ingredients) amount to 20 drachmae; this must be added to them: Saffron from Cilicia, 4 drachmae; with this write on whatever (surface) you want, on vessels, on paper as well as on stones and wood.

When read in parallel, these three recipes provide an almost complete repertoire, so to speak, of the "commonest non-metallic ingredients for supplementing metal powder," as Alexander wrote in her overview of the recipes for metallic inks in the Western Middle Ages (an overview that did not consider the Syriac evidence):[65] gold-coloured orpiment (a shiny yellow arsenic sulfide); pomegranate; bile of various animals; yellow plants such as celandine juice (often substituted with a plant called W'RŠ' in the Syriac recipes); and saffron. The same set of ingredients is used in recipe 72 of the Leiden Papyrus

63 See above, p. 90.

64 When compared with similar descriptions of the same technique preserved in Greek and Latin (see below), the text seems to be incomplete (perhaps a few words were lost in the transmission of the recipe). We would have expected to find a reference to 5 parts of egg white here.

65 Alexander, *Medieval Recipes* (cit. note 53), p. 42.

as well as in recipes 47 and 50 of the *Mappae clavicula*. Texts and translations of these three recipes are provided below:

Leiden Papyrus, **rec. 72**[66]
Ἄλλη. Χρυσογραφία χωρὶς χρυσοῦ· ἐλυδρίου μέ(ρος) α′, ῥητίνης καθαρᾶς μέ(ρος) α′, ἀρσενικοῦ χρυσίζοντος μέ(ρος) α′ ὅ ἐστιν σχιστόν, κόμμεως καθαροῦ, χολῆς χελώνης μέ(ρος) α′, ᾠῶν τοῦ ὑγροῦ μέ(ρη) ε′, ἤτω δὲ τῶν ξηρῶν πάντων ἡ ὁλκὴ Σ κ′, εἶτα ἐπέμβαλε τούτοις κρόκου κιλικίου Σ δ′. Ποίει δὲ οὐ μόνον ἐπὶ χάρτου ἢ διφθέρας, ἀλλὰ καὶ ἐπὶ μαρμάρου ἐστιλβωμένου καὶ ἐάν τι ἄλλο καλὸν θέλῃς ὑποζωγραφῆσαι καὶ ποιῆσαι χρυσοειδές.

Another (recipe). To write in letters of gold without gold. Celandine, 1 part; pure resin, 1 part; golden coloured orpiment, the one that is scissile, 1 part; pure gum; bile of tortoise, 1 part; the liquid part of eggs, 5 parts; take 20 staters by weight of all these materials dried; then throw in 4 staters of saffron of Cilicia. Can be used not only on papyrus or parchment, but also upon highly polished marble, or also when you wish to make a beautiful design upon some other object and give it the appearance of gold.[67]

Mappae clavicula, **rec. 47**[68]
Auri alia scriptio sine auro. Elydrii partem I, resinae frixae partem I, ovorum V humores, gummi puri partem I, auripigmenti scissilis partem I, fellis testudinis partem I. Sit autem eorum id est tunsorum omnium pondus ad dragmas XX. Deinde adicias croci ciliciensis dragmas II. Fac autem hoc non solum in cartis et in membranis, verum etiam in marmore et in vitro.

Another writing in gold without gold. Celandine, 1 part; broken resin, 1 part; the white of 5 eggs; pure gum, 1 part; scissile orpiment, 1 (part); gall of a tortoise, 1 (part). The weight of them all, after they have been pounded, should be about 20 drachmae. Then add 2 drachmae of Cilician saffron. This works not only on papyrus and parchment, but also on marble and on glass.[69]

66 Halleux, *Papyrus* (cit. note 2), p. 101. See also the recipe 56 of the Leiden papyrus, which records a list of similar ingredients: Halleux, *Papyrus* (cit. note 2), p. 98.
67 Translation (slightly modified) by Caley, *The Leiden Papyrus* (cit. note 50), p. 1159.
68 Baroni et al., *Mappae clavicula* (cit. note 50), p. 108.
69 Translation based on Cyril Stanley Smith and John G. Hawthorne, "*Mappae Clavicula.* A Little Key to the World of Medieval Techniques," *Transactions of the American Philosophical Society*, 1974, 64,4:1–128 (p. 34).

Mappae clavicula, **rec.** 50[70]
Aurei coloris scriptura in cartis, in marmore et vitro ut videatur de auro. Elydrii partem I, auripigmenti partem I, fellis testudinis partem I, aluminis scissilis partem I et de corio mali punici quod intus est aurei coloris I, gummi I, ova V. Sit autem eorum pondus dragmas IX et croci dragmas II.

Gold-coloured writing on papyrus, marble, and glass, so that is seems to be made of gold. Celadine, 1 part, orpiment, 1 part, gall of a tortoise, 1 part, scissile alum, 1 (part) and 1 part of the skin of a pomegranate that is gold-coloured inside, gum, 1 part, 5 eggs. The weight of all these should be 9 drachmae, and 2 drachmae of saffron.[71]

All these recipes, except for **R8** of the Cambridge manuscript, emphasize the multiple applications of the produced ink, which could be used to write on different supports, from papyrus and parchment to metal, marble, wood, stone, and glass. In order to better compare the substances involved in the described procedures, I have listed the ingredients in Table 5.1.

The Syriac recipe included in Democritus' book overlaps in many respects with Leiden Papyrus, rec. 72, and *Mappae clavicula* (hereafter *MC*) rec. 47. Both the ingredients and their proportions are almost the same, and some variations can probably be explained with textual arguments. As for the amount of eggs, the Leiden Papyrus and *MC* 47 specifies taking 5 parts, while Democritus' Syriac text mentions the same quantity for the bile of a calf, but omits any mention of eggs: we cannot exclude that the term 'eggs' was originally in the Syriac recipe (or in its source), but was later omitted by a scribe who copied the text.[72] Likewise, the variation between "bile of a tortoise" (both in the Leiden Papyrus and in *MC* 47) and "bile of a calf" (Democritus' recipe) can be explained by considering the transmission of the Syriac text: in Syriac, in fact, the terms for the two animals are very similar — ܓܠܐ (*galo*), 'tortoise' and ܥܓܠܐ (*ʿeglo*), 'calf' — and they can be easily interchanged. Philologists and critical editors, however, should be very cautious in handling similar texts, since lexical variations can also point to subtle changes in the selection of the ingredients. Indeed, the Leiden Papyrus (rec. 61) specifies that very bitter bile of a calf (μοσχεία χολὴ κατάπικρος) could be used instead of tortoise bile (χολὴ χελώνης) in the making of a golden ink.[73]

---

70 Baroni et al., *Mappae clavicula* (cit. note 50), p. 110.
71 Translation based on Smith and Hawthorne, *Mappae Clavicula* (cit. note 69), p. 35.
72 See also **R7**, where 5 eggs are recorded.
73 Halleux, *Papyrus* (cit. note 2), p. 99; see also Alexander, *Medieval Recipes* (cit. note 53), p. 42.

TABLE 5.1 Ingredients of the golden ink according to ancient texts

	R7	R8	Democritus	Leiden 72	*MC* 47	*MC* 50
Plants						
plant dye *W*ꜣ*RŠ*ꜣ	✓ (1 part)	✓ (1 part)				
plant dye 'celandine'			✓ (1 part)	✓ (1 part)	✓ (1 part)	✓ (1 part)
pomegranate (inner part)	✓ (1 part)					✓ (1 part)
saffron	✓ (1 part)		✓ (4 drach.)	✓ (4 stat.)	✓ (2 drach.)	✓ (2 drach.)
resin				✓ (1 part)	✓ (1 part)	
gum		✓ (1 part)		✓	✓ (1 part)	✓ (1 part)
Minerals						
orpiment	✓ (1 part)	✓ (1 part)	✓ (1 part)	✓ (1 part)	✓ (1 part)	✓ (1 part)
verdigris		✓ (1 part)				
alum	✓ (1 part)					✓ (1 part)
earth of Samos	✓ (1 part)					
metallic leaf		✓				
Animal products						
bile of a calf		✓ (1 part)	✓ (5 parts)			
bile of a tortoise	✓ (1 part)			✓ (1 part)	✓ (1 part)	✓ (1 part)
eggs	✓ (5 eggs)			✓ (5 parts)	✓ (5 eggs)	✓ (5 eggs)

A clear similarity in the ingredients and their quantities is also detectable between the Syriac recipe **R7** and *MC* 50: as for the dye plant used in the process, the Syriac text prescribes using *W'RŠ'* rather than celandine,[74] and it also includes the earth of Samos, which is not mentioned in the *Mappae clavicula*.

If we go back to the recipe book on ink making preserved by the Cambridge manuscript, we must note that recipes 8–9, which do not require the use of precious metals, are followed by a cluster of recipes that describe various treatments of gold for preparing inks and paints:

(R9) Fol. 2v17–3r11 (golden ink with the precious metal)
Inc.: ܣܝܪ. ܫܦܘܟ ܒܐܘܢ. ܘܐܦܪܘܕܝܗܘܢ, "*Sir*. Pour lead and copper etc."

(R10) Fol. 3r11–3v2 (golden ink with the precious metal)
ܣܝܪ ܕܢܟܬܘܒ ܐܝܟ ܡܢ ܕܗܒܐ. ܐܪܣܢܝܩܘܢ ܕܫܦܝܪ ܡ̈ܢܘܢ ܬܪ̈ܬܝܢ. ܘܐܪܫܐ ܡܢܬܐ
ܚܕܐ. ܡܪܬܟܐ ܕܕܗܒܢ ܡܢܬܐ ܚܕܐ. ܫܚܘܩ ܘܦܬܟ ܒܡ̈ܝܐ. ܘܒܬܪܟܢ ܣܒ ܩ̈ܦܠܐ ܒܥܣܪ̈ܝܢ
ܘܐܪܒܥܐ. ܘܐܪܡܐ ܚܕܐ ܡܢ ܐܪܒܥ ܦܬܟܐ. ܘܫܚܘܩ ܐܟܚܕܐ ܒܡܘܪܟܢ ܕܕܟܝܐ ܘܟܕ ܪܡܐ
ܐܝܟ ܡܟܝܠ ܡܠܚܐ. ܘܟܕ ܕܐܬܫܚܩ ܫܦܝܪ: ܐܪܡܐ ܡ̈ܝܐ ܥܕܡܐ ܕܦܐܫ ܗܘ ܕܗܒܐ
ܒܠܚܘܕܘܗܝ ܕܟܝܐ. ܘܒܬܪܟܢ ܐܪܡܐ ܡܢܗ ܕܫܚܠܐ ܗܘ ܡܢܗ ܕܦܬܟ. ܘܩܘܡܝ ܕܪ̈ܥܝܥܐ
ܩܠܝܠ. ܟܕ ܪܡܐ ܐܢܬ ܥܠܘܗܝ ܩܠܝܠ ܡܢ ܟܘܪܟܡܐ. ܘܫܚܘܩ ܥܕܡܐ ܕܥܒܐ ܐܝܟ
ܕܝܘܬܐ ܘܫܝܡܝܗܝ ܒܡܐܢܐ ܕܢܚܫܐ. ܘܡܐ ܕܟܬܒ ܐܢܬ ܒܗ ܨܒܘܥ ܩܢܝܐ ܒܙܪ̈ܦܐ[75] ܪܛܝܒ
ܘܟܬܘܒ. ܘܡܐ ܕܝܒܫ ܟܠܘܣ ܒܡܫܟܐ ܕܚܙܝܪܐ.

Sir. To write with gold. Good orpiment (= Gr. ἀρσενικόν), 2 parts; *W'RŠ'* (a dye plant), 1 part; litharge that we have gilded, 1 part; grind and mix with water. Then take 24 leaves (of gold?) and add a quarter of the mixture. Grind together in a clean mortar by adding a bit of salt. When it is well ground, add water until only pure gold remains. Then add what is left of the mixture and a bit of broken gum, by pouring a bit of water of saffron over it. Then you grind until it thickens like an ink (*dyuto*) and put it in a copper vessel. When you write with it, soak a reed ⟨in⟩ liquid alum, and write. When it gets dry, polish with pork rind.

This detailed recipe singles out the different passages of the procedure, at the same time specifying the right proportions of the various substances added to the preparation of the golden ink. Quite striking is the mention of 24 leaves of gold (l. 2), which seems to imply the use of a significant amount of the precious

74 The same substitution can be observed in **R10** (see below).
75 I added the prep. ܒ; the MS simply reads ܙܪ̈ܦܐ. See also below, **R13**.

metal. The same datum, however, appear to be confirmed by the Latin version of this recipe (clearly based on a common Greek source), which is included in the *Mappae clavicula* (rec. 38):[76]

> *Auripigmenti scissilis partes II, elydrii partem I, spumae argenti cuius color sit aureus partem I. Haec, cum triveris, diffunde in vase. Postea accipe laminas aureas XXIIII ad quartam, quantum voles ex his tere in mortario mundo medicinali. Adice sal modicum et, cum tibi apparverit ut arena diligenter trita, adice aquam puram et tere et ablue, ita ut frequenter aquam effundas et aliam infundas donec tibi aurum purum appareat. Tunc adice de suprascripto medicamine quod sufficiat et modicum gummi triti, ita ut non sit glutinosum. Instilla destillationem croci et omnia simul tere, ut sit quasi atramenti pinguedo, et recipe aut in concam aut in vitreum vas. Cumque scribere vis, primum ungue cannam liquido alumine et tunc in aurum intingue et scribe et, cum siccaverit, dente frica diligenter.*
>
> Scissile orpiment, 2 parts, celandine, 1 part, litharge, whose colour must be golden, 1 part. After grinding them, pour them into a pot. Then take as fourth (part) 24 gold leaves, and grind as much as you want of these in a clean pharmacist's mortar. Add a bit of salt and, when it looks like well-ground sand, add fresh water, grind and wash it off, so that you continually pour out the water and add new water until you see that the gold is pure. Then add a sufficient amount of the above-mentioned drugs and a little ground gum, so that it does not get sticky. Drip onto it an extract of saffron and grind everything together, so that it has the consistency of ink. Place it in a shell or a glass pot, and when you want to write, first smear the reed pen with moist alum, then dip it in the gold and write. When it gets dry, rub it thoroughly with a tooth.[77]

Only a few differences between the Latin and the Syriac text are detectable. As already noted in the comparison between R7 and *MC* 50, the dye plant called *W'RŠ'* is used in the Syriac text (R10) instead of celandine (*elydrium* = Gr. ἐλύδριον), which is prescribed in the Latin recipe (*MC* 38). Moreover, according to the latter, the ink was to be placed in a shell or a glass pot rather than in a copper vessel as in the Syriac recipe. Here, pig skin is used to polish the written letters, while the Latin text mentions using a tooth.

---

76 Baroni et al., *Mappae clavicula* (cit. note 50), p. 100.

77 Translation based on Smith and Hawthorne, *Mappae Clavicula* (cit. note 69), p. 34.

Another six recipes for golden inks follow on from the above-mentioned formula (**R10**) in the Cambridge manuscript, some of which again reveal similarities with the *Mappae clavicula*.

(R11) fol. 3v2–12. (gold ink)
Inc.: ܐܚܪܢܐ ܕܫܦܝܪ, "Another beautiful (recipe)."

(R12) fol. 3v12–19 (gold ink)
Inc.: ܣܝܪ ܕܢܕܗܒ ܐܝܟ ܦܫܝܩܐܝܬ, "*sir*. To gild easily."

(R13) fol. 3v20–4r3 (gold ink with the precious metal)
ܕܢܟܬܘܒ ܐܝܟ ܒܕܗܒܐ. ܣܒ ܐܒܪܐ ܘܫܘܚ ܙ̈ܒܢܝܢ ܣ̈ܓܝܐܢ ܟܕ ܨܒܥܬ ܠܗ ܒܡ̈ܝܐ ܩܪ̈ܝܪܐ. ܘܒܬܪܟܢ ܫܚܘܩ ܠܕܗܒܐ ܥܡ ܢܢܐ ܕܟܝܐ. ܥܕܡܐ ܕܡܬܪܦܐ. ܘܗܝܕܝܢ ܐܪܡܐ ܒܗ ܡ̈ܝ ܩܘܡܝ ܘܟܬܘܒ. ܟܕ ܨܒܥܬ ܩܢܝܐ ܒܨܪܘܦܐ[78] ܪܛܝܒܐ.

To write with gold. Take lead and melt it several times, while quenching it in cold water. Then grind gold with pure mercury (*ʿnono*, lit. 'cloud') until it softens. Then, add water of gum in it and write, by dipping a reed ⟨in⟩ moist alum.

(R14) fol. 4r3–15 (Gold paint/ink)
Inc.: ܣܝܪ ܕܢܗܘܐ ܕܢܚܙܐ ܟܠ ܡܐܢܐ ܐܝܟ ܕܗܒܐ, "*sir*, so that a vessel looks like gold."

The same technique described in **R13** also occurs in the *Mappae clavicula*. The various manuscripts of the Latin compilation, however, preserve different versions of this recipe with different degrees of complexity in terms of explaining the procedure. Below, I have copied the version transmitted by MS Corning Museum of Glass, Philipps 3713, which shows close similarities with the Syriac version:

Mappae clavicula, **rec. 44**[79]
Aliter. Plumbum conflas frequenter et intinguis in aquam frigidam et tunc conflabis aurum et restringuis in predicta aqua et fit fragile. Deinde teris diligenter aurum cum argento vivo, ipsam autem fecem cum quo scis diligenter purgas et misces gummi liquidum et scribe, ante in alumine liquido calamum tingue. Et sale et aceto purges alumen.

78 I added the prep. ܒ; the MS simply reads ܨܪܘܦܐ. See also above, **R10**.
79 Baroni et al., *Mappae clavicula* (cit. note 50), p. 106.

Another way. You melt lead several times and quench it in cold water; then you will melt some gold and quench it in the above-mentioned water; it becomes brittle. Then you grind the gold thoroughly with quicksilver. You clean the dregs with the substance that you know and mix liquid gum, and write, after you have dipped the reed in liquid alum. Clean the alum with salt and vinegar.[80]

As already noted with reference to **R6**, the addition of lead makes the gold brittle and facilitates the grinding process. In this case, the procedure is combined with the standard use of mercury for reducing gold to powder. On the basis of laboratory experiments, Alexander explains that, if added in the right proportion, "gold leaf readily amalgamates with mercury to form a pulverulent mass."[81]

The following cluster of two recipes (**R15–16**) corresponds to two texts that also appear consecutively in the *Mappae clavicula* (MC 34–35), further confirming the close link between the Syriac and the Latin compilations, which probably drew on common sources.[82]

(R15) fol. 4r15–18 (Gold paint/ink).
ܣܝܪ ܕܢܥܒܕ ܐܝܟ ܡܫܚܐ ܕܕܗܒ. ܣܒ ܩܢܘܒܪܝܣ ܘܚܠܐ ܕܛܘܪܐ ܘܫܘܦܐ ܕܕܗܒܐ
ܘܙܘܦܐ ܘܚܠܐ ܘܫܘܦ ܐܟܚܕܐ ܘܒܫܠ ܒܡܐܢܐ ܕܢܚܫܐ ܘܐܪܕܝܘܗܝ ܥܕ ܕܢܫܬܪܐ.

Sir. To make gold liquid. Take cinnabar, sand of a mountain, filings of gold, alum, and vinegar, grind them together, boil in a vessel of copper and stir until it is melted.

Mappae clavicula, **rec. 34.**[83]
Aurum liquidum facere. Minium et arenam montanam, auri limaturam et alumen cum aceto simul tere et coque in vase aereo et commove. Huius scripturae color per annos durat.

80 Translation based on Smith and Hawthorne, *Mappae Clavicula* (cit. note 69), p. 34.
81 Alexander, *Medieval Recipes* (cit. note 53), p. 38.
82 We must note that, according to Baroni's analysis, the second part of the Syriac treatise includes three recipes that appear in the same order in the *Mappae clavicula*: Mm. 6.29, fol. 7r10–21 *How (to treat) silver, copper or gold for writing* = MC 122; Mm. 6.29, fol. 7r21–7v2 *To write letters on Cyprian copper* = MC 123; Mm. 6.29, fol. 7v3–10 *To write black letters on a copper vessel, which cannot be erased* = MC 124. See Baroni et al., *Mappae clavicula* (cit. note 50), pp. 157–158 and 240.
83 Baroni et al., *Mappae clavicula* (cit. note 50), p. 96.

Make the gold liquid. Cinnabar and mountain sand, filings of gold and alum with vinegar; grind together, boil in a copper vessel and stir. The colour of this writing lasts for years.

(R16) fol. 4r19–4v7 (Gold paint/ink)

ܕܢܗܘܐ ܐܝܟ ܡܝܐ ܕܗܒܐ ܐܝܟ ܕܢܟܬܘܒ ܒܗ. ܢܛܪܘܢ ܣܘܡܩܐ ܙܘܙܐ ܬܪܬܝܢ. ܩܝܢܢܒܪܝܣ
ܙܘܙܐ ܬܠܬܐ. ܚܠܛ ܘܫܚܘܩ ܒܚܠܐ. ܘܐܪܡܐ ܙܥܘܪ ܡܢ ܫܒܐ ܘܫܒܘܩ ܢܒܫ. ܘܒܬܪܟܢ
ܫܚܘܩ ܘܦܠܓ. ܘܣܒ ܫܘܦܐ ܕܕܗܒܐ ܙܘܙܐ ܬܪܬܝܢ. ܘܐܪܣܢܝܩܘܢ ܕܡܬܦܠܓ ܙܘܙܐ ܚܕ. ܘܚܠܛ
ܐܟܚܕܐ ܘܫܚܘܩ. ܟܕ ܪܡܐ ܐܢܬ ܡܢ ܡܝܐ ܕܩܘܡܝ ܕܟܝܐ ܘܣܒ ܘܛܒܘܥ ܟܠ ܡܐ ܕܨܒܐ.
ܘܫܒܘܩ ܠܬܪܝܢ ܝܘܡܝܢ ܘܢܒܫ. [...].

Making gold liquid to write with it. Red soda (= Gr. νίτρον), 2 coins (corresponding to the Greek drachma); cinnabar (= Gr. κιννάβαρις), 3 coins; mix and grind with vinegar, add a bit of alum, and let it dry out. Then grind and take it apart (?). Take filings of gold, 2 coins; scissile orpiment, 1 coin; mix together and grind, by adding water of pure gum; take it and seal what you want. Leave it for two days and it gets dry. Etc.[84]

Mappae clavicula, **rec. 35**[85]

Aurum mollire ut in eum sigillum figas. Nitri rufi dragmas II, minii dragmas III permisce et tere cum aceto et adice aluminis modicum et dimitte ut siccetur. Deinde tere et repone. Et sume auri limaturam ad dimidium oboli et auripigmenti scissilis dragmam I, misce omnia et tere et diffunde gummi puram in aquam infusum. Tolle et signa quod velis, sive epistulam, sive tabulam, et dimitte biduo usque coagulatur sigillum.

Mix two drachmae of red soda (*nitrum*) and 3 drachmae of cinnabar. Grind with vinegar, add a little alum and leave it to dry. Then grind it and leave it aside. Take about half an obol of gold filings and 1 drachma of scissile orpiment, mix them all, grind them and pour over an infusion of gum in pure water. Take it out and seal what you want, whether a letter or a tablet. Leave it for two days until the seal becomes hard.[86]

---

84 I have omitted an additional remark at the end of the recipe, which seems to explain how to use the ink for several applications.

85 Baroni et al., *Mappae clavicula* (cit. note 50), p. 96.

86 Translation based on Smith and Hawthorne, *Mappae Clavicula* (cit. note 69), p. 33.

5 Concluding Remarks

A galaxy of recipes on ink making emerges from the comparison of the instructions preserved by multiple traditions, which, despite their own peculiarities and different ramifications, appear to be firmly rooted in a shared and rich Late Antique material. In particular, the close similarities between Syriac and Latin recipes cannot be properly assessed without supposing common Greek sources that were independently translated into either language by scholars working in distinct cultural settings and operating in different periods. Traces of this Greek Late Antique heritage can be detected in Graeco-Egyptian papyri, such as the Leiden Papyrus, which represents a crucial source for reconstructing ancient procedures used to prepare gold or silver inks and paints. Other relics of this ancient technology are detectable in the collection of Greek alchemical treatises transmitted by Byzantine manuscripts, which share various features with the Syriac sources under examination.

On the other hand, the Syriac manuscript Mm. 6.29 includes a variety of procedures for the making of metallic inks, which substantially enriches the scanty information that can be extracted from the material preserved in Graeco-Egyptian papyri as well as in Byzantine alchemical sources. Gold and silver inks were important tools belonging to a broader umbrella of techniques that aimed at changing the colours of different materials, from metals and stones to papyrus and parchment. As already seen, commenting on the recipes included in the Leiden Papyrus, Robert Halleux rightly emphasized that: "La composition des encres est rigouresement parallèle à celle des dorures et des vernis."[87] After all, Late Antique alchemical theories and practices developed around the effort to select and conceptualize a broad spectrum of techniques dealing with chromatic transformations. The Syriac tradition testifies to the centrality of colours and dyes in the ancient alchemical discourse, which included the making of inks among the areas of expertise to be explored, organized, and handed down over centuries.

---

87 Halleux, *Papyrus* (cit. note 2), p. 42.

CHAPTER 6

The Literary Dimension and Life of Arabic Treatises on Ink Making

Sara Fani

Abstract

In Arabic literary tradition, single ink recipes are scattered in works of different genres, from the alchemical and medical, to those related to calligraphy and penmanship and dedicated to the class of the *kuttāb*. A handful of treatises stand out for their collection of a significant number of recipes, organized in categories and juxtaposed with other textual sections on different technical crafts. Ranging from the 9th to the 17th centuries and from al-Andalus to Yemen, they show a great fluidity in their transmission, fostered by their fragmented structure in short textual units. This contribution presents a series of case studies highlighting the modalities of formation of these compilations and the literary elements that emerge alongside their technical content. This can only be retrieved and properly interpreted by taking into account their literary dimension, which reflects the cultural context in which these treatises have been generated.

Keywords

Arabic literature – compilations of recipes – textual transmission

1 Introduction

The present enquiry about inks in the Arabo-Islamic manuscript tradition is focused on compilations of technical-artistic recipes on ink making in the Arabic language.* The study considers only those technical works arranged as collections of recipes on ink making that were disseminated and developed as

* The last reviews of this contribution occurred after a change of academic affiliation to the University of Naples "L'Orientale," for the *EuQu* ERC project (*The European Qur'an. Islamic Scripture in European Culture and Religion 1150–1850*).

DOI:10.1163/9789004444805_008

identifiable textual traditions under the name of an author.[1] This selection is based upon the original intention of my research, i.e. mapping the procedures and materials employed in ink making in different ages and regions of the Islamic world. For this reason, anonymous works are only considered from a comparative perspective, and not as meaningful sources for a technical history of ink making.[2] This was the perspective that initially informed my approach to this literary production, and which became the basis for further enquires and chemical experimentations based on the technical content of the texts examined.[3] Nonetheless, the textual and literary presentation of the technical content are crucial elements for the correct interpretation of the documents. This present contribution will therefore focus on some features of the works in question that demonstrate their literary dimension and aspects of their textual transmission.

2 Ink Recipes in the Arabic Literary Tradition

Occurrences of single ink recipes within the Arabic literary tradition are attested in various sources. The first ink recipe in Arabic is transmitted in an alchemical treatise, the *Kitāb al-ḫawāṣṣ* ('The book of the occult properties') ascribed to Ǧābir ibn Ḥayyān (8th c.), one of the main representatives of the

1 The specific aspects that form the focus of this chapter have been developed from my PhD research on the subject. See Sara Fani, Le arti del libro secondo le fonti arabe originali. I ricettari arabi per la fabbricazione degli inchiostri (sec. ix–xiii): loro importanza per una corretta valutazione e conservazione del patrimonio manoscritto (PhD Diss., University of Naples "L'Orientale," 2013).

2 Anonymous sources are edited in ʿAbd al-Laṭīf al-Rāwī, ʿAbd al-Ilāh Nabhān, "Risāla fī ṣināʿat al-kitāba," *Maǧallat Maǧmaʿ al-luġa al-ʿarabiyya bi-Dimašq = Revue de l'Académie Arabe de Damas*, 1987, 62/4: 760–795; and Idem, "Risāla fī ṣināʿat al-kitāba (al-qism al-ṯānī)," *Maǧallat Maǧmaʿ al-luġa al-ʿarabiyya bi-Dimašq = Revue de l'Académie Arabe de Damas*, 1988, 63/1: 50–65; Barwīn Badrī Tawfīq, "Risālatān fī ṣināʿat al-maḫṭūṭ al-ʿarabī," *al-Mawrid*, 1985, 14/4:275–286; Eugenio Griffini, "Nuovi testi arabo-siculi," in *Centenario della nascita di Michele Amari: scritti di filologia e storia araba, di geografia, storia, diritto della Sicilia medievale, studi bizantini e giudaici relativi all'Italia meridionale nel Medio Evo, documenti sulle relazioni fra gli Stati italiani ed il Levante*, 2 vols (Palermo: Virzì, 1910), vol. I, pp. 443–448; the latter is translated in Cristina La Rosa, "Alcune ricette per la preparazione degli inchiostri *ḥibr* e *midād* tratte dal Libro del Siciliano: Traduzione del testo e osservazioni," in *Islamic Sicily: Philological and Literary Essays*, edited by Mirella Cassarino (*Quaderni di Studi Arabi n.s.*, 2015, 10), pp. 173–190.

3 See the recently discussed PhD thesis by Claudia Colini, From Recipes to Material Analysis: The Arabic Tradition of Black Inks and Paper Coatings (9th to 20th century) (PhD Diss., University of Hamburg, 2018).

Arab alchemical tradition.[4] Ink recipes are also transmitted in treatises on calligraphy being an essential part of the equipment for scribes and copyists; for example, an ink recipe is recorded by the famous calligrapher Ibn Muqla (9th–10th c.) in his work *Risāla fī al-ḫaṭṭ wa-l-qalam* ('Short treatise on calligraphy and reed pen').[5] Moreover, in a commentary on the poetical work ascribed to another famous calligrapher, Ibn al-Bawwāb (10th–11th c.), entitled *Rāʾiyya fī al-ḫaṭṭ* ('The poem in rhyme *Rāʾ* on calligraphy'), two recipes by Ibn al-Bawwāb and others by his commentators are reported.[6] Ink recipes are also attested in manuals compiled by and for the class of the *kuttāb* (sing. *kātib*, court chancellors, clerks or secretaries) such as *al-Risāla al-ʿaḏrāʾ* ('The virgin epistle'), a short treatise on epistolary prose by al-Šaybānī (second half of the 9th c.),[7] or the later *Ṣubḥ al-aʿšā fī ṣināʿat al-inšāʾ* ('Dawn for the blind in the craft of composition') by Abū al-ʿAbbās al-Qalqašandī (756–821/1355–1418).[8]

Another genre in which ink recipes appear deserves to be mentioned for the peculiar social and cultural context in which it was produced; it was in fact,

---

4 Manfred Ullmann, *Die Natur- und Geheimwissenschaften in Islam* (Leiden: Brill, 1972), p. 408; Armin Schopen, *Tinten und Tuschen des Arabisch-islamischen Mittelalters: Dokumentation, Analyse, Rekonstruktion; ein Beitrag zur materiellen Kultur des Vorderen Orients* (Göttingen: Vandenhoeck & Ruprecht, 2006), pp. 36–37, n. 2. For an excursus of the specific genre of books of properties (*manāfiʿ* and *ḫawāṣṣ*), see Lucia Raggetti, "The 'Science of Properties' and its Transmission," in *In the Wake of the Compendia. Infrastructural Contexts and the Licensing of Empiricism in Ancient and Medieval Mesopotamia*, edited by J. Cale Johnson (Berlin/Boston: De Gruyter, 2015), pp. 159–176 and Eadem, *ʿĪsā ibn ʿAlī's Book on the Useful Properties of Animal Parts. Edition, Translation and Study of a Fluid Tradition* (Berlin/Boston: De Gruyter, 2018), pp. xiv–xvi.

5 Dominique Sourdel, "Ibn Mukla," in *Encyclopaedia of Islam, New Edition*, ed. by Peri J. Bearman, Thierry Bianquis, Clifford E. Bosworth, Emeri J. van Donzel, Wolfhart P. Heinrichs vol. III pp. 886b–887a; Abū ʿAlī Muḥammad b. ʿAlī Ibn Muqla, "Naṣṣ risālat Ibn Muqla fī al-ḫaṭṭ wa-l-qalam," in *Ibn Muqla, ḫaṭṭāṭan wa-adīban wa-insānan*, edited by Hilāl Nāǧī (Baġdād: Dār al-šuʾūn al-ṯaqāfiyya al-ʿāmma, 1991), pp. 115–116.

6 Janine Sourdel-Thomine, "Ibn al-Bawwāb," in *Encyclopaedia of Islam, New Edition* (cit. note 5), vol. III pp. 736b–737a; David James, "The Commentaries of Ibn al-Baṣīṣ and Ibn al-Waḥīd on Ibn al-Bawwāb's *Ode on the Art of Calligraphy* (*Rāʾiyyah fī al-ḫaṭṭ*)," in *Back to the Sources. Biblical and Near Eastern Studies in Honour of Dermot Ryan*, edited by K.J. Cathcart and J.F. Healey (Dublin: Glendale, 1989), pp. 172–173, 188–189.

7 Ibrāhīm Ibn al-Mudabbir, *Risālat al-ʿaḏrāʾ = Etude critique sur La lettre vierge d'Ibn el-Mudabber*, ed. by Zākī Mubārak (al-Qāhira: Dār al-kutub al-miṣriyya / Paris: Maisonneuve Frères, 1350/1931). In this edition, the work was ascribed to Ibrāhīm Ibn al-Mudabbir (m. 279/893) who was its dedicatee (cf. Dominique Sourdel, "Le 'Livre des secrétaires' de ʿAbd Allāh al-Baghdādī," *Bulletin d'Etudes Orientales*, 1952–1954, 14:115–153, p. 116, note 2).

8 Abū al-ʿAbbās Aḥmad b. ʿAlī al-Qalqašandī, *Ṣubḥ al-aʿšā fī ṣināʿat al-inšāʾ*, 14 vols (al-Qāhira: Maṭbaʿa al-amīriyya, 1918), Vol. 2, p. 465.

within the milieu of the Banū Sāsān that Muḥammad ibn Abī Bakr al-Zarḫūrī (9th/15th c.) compiled his *Kitāb zahr al-basātīn* ('Book of the flowers of the gardens'). The recipes transmitted by this source were part of the corpus emanating from an urban Islamic "underworld" comprised of magicians, fortune tellers, herbalists, illusionists, etc. that had the specific aim of amazing the public with astonishing effects.[9]

A few procedures related to the production of black and coloured inks are also attested in treatises related to medicinal, or better, natural remedies; for example, *Al-urǧūza al-faṣīḥa fī al-aʿmāl al-ṣaḥīḥa* ('The eloquent poem in *Raǧaz* verses on the right crafts') by Abū Bakr Ibn Yaḥyā al-Kātib al-Ḫarrāṭ,[10] which features a number of ink recipes grouped by procedures for different daily personal and domestic concerns, such as removing stains from clothes, pest control (insects and other animals), medical preparations and dietary indications. This is not the only example in which the recipes are written in verse, as attested by its title (*urǧūza*),[11] and it is interesting to note that, in this case, it is the author himself who explains the advantages of a rhymed composition: because it is "easy to remember by heart and to transmit" (*sahl fī ḥifẓihī wa-naqlihī*).[12]

A similar work is attested in a manuscript collected, and possibly commissioned, by the famous traveller James Bruce (1730–1794). The manuscript, now kept in the Bodleian Library in Oxford (Ms Bruce 45, ff. 10v–25v),[13] is a codicological curiosity as it was copied onto parchment in the 18th century, in Ethiopia (Gondar), where Arabo-Islamic manuscripts were usually transmitted on paper. The work is entitled *Kitāb al-Azraq fī al-ṭibb* ('Al-Azraq's

9 Lucia Raggetti, "*Cum Grano Salis*. Some Arabic Ink Recipes in Their Historical and Literary Context," *Journal of Islamic Manuscripts*, 2017, 7:294–338, pp. 328–334 and Clifford E. Bosworth, *The Mediaeval Islamic Underworld: The Banū Sāsān in Arabic Society and Literature*, 2 vols (Leiden: Brill, 1976).

10 Carl Brockelmann, *Geschichte der Arabischen Litteratur. Zweiter Supplementband* (Leiden: Brill, 1938), p. 1029. The work is attested in a manuscript in Gotha (ms. orient. A 1327) and in two manuscripts in Paris: BnF Ar. 2776, ff. 64r-71r and BnF 6844, ff. 134v–141r, copied at the end of Ḏū al-Ḥiǧǧa 986 H/February 1579 CE.

11 One case is represented by the above-mentioned poem by Ibn al-Bawwāb, for others cf. Colini, *From recipes* (cit. note 3), pp. 57–58.

12 See MS Paris, BnF Ar. 2776, f. 64v, ll. 7–8. For similar mnemotechnical features in Western treatises on colour making, see Sandro Baroni, Paola Travaglio, "Mnemotecnica e aspetti di oralità nei ricettari di tecniche dell'arte e dell'artigianato," *Studi di Memofonte*, 2016, 16:114–129.

13 Emilie Savage-Smith, *A New Catalogue of Arabic Manuscripts in the Bodleian Library, University of Oxford. Volume I: Medicine* (Oxford: Oxford University Press, 2011), pp. 429–430, entry n. 107. See also the database of the ERC project *Islam in the Horn of Africa*, where the manuscript has been described by the present author (<http://islhornafr.tors.sc.ku.dk/backend/texts/5143>, last accessed February 2019).

book on medicine'), attributed to Aḥmad al-ʿAbdallāwī al-Maḥallawī (fl. 1770), and seems to be an excerpt of a Yemeni medical treatise by Ibrāhīm b. ʿAbd al-Raḥmān b. Abī Bakr (b.) al-Azraq (fl. 890/1485), which is kept in the same collection, but copied — on paper — some decades before Bruce's stay in the region.[14] The few recipes transmitted in these manuscripts are dedicated to coloured inks (*līqa*).

3 Compilations of Ink Making Recipes

A small number of treatises developed sections dedicated to ink making, describing a considerable number of recipes and organizing them in groups according to their characteristics; these chapters are juxtaposed with sections related to other kinds of recipes or instructions for different technical craftmanships. The present analysis primarily focuses on compilations produced until the 13th century, a time span that represents the period commonly identified as the Golden Age of classical Arabo-Islamic civilization and its literary productions. Nonetheless, later compilations on inks show a degree of originality and deserve specific attention, also with reference and in comparison to the more ancient traditions.[15] A brief description of the sources will be given in order to attest their available editions or witnesses and their general content.[16]

The *Zīnat al-kataba* ('The ornament of the scribes'), is the oldest known technical treatise dealing with sets of ink. It is ascribed to Abū Bakr Muḥammad b. Zakariyyāʾ al-Rāzī (d. 313/925), one of the most famous medieval physicians and alchemists, born in Rayy where he was educated according to the Greek scientific tradition. The above-mentioned short essay is known only through one witness (dated 907/1502), found in the Dār al-Kutub, Cairo, in 2010 and now published.[17] The short treatise focuses solely on the preparation of black

---

14 MS Oxford, Bodleian Library, Bruce 7 described in Savage-Smith, *A New Catalogue* (cit. note 13), pp. 738–741, entry n. 218.

15 Among the later sources are the treatises by al-Dimašqī (10th/16th c.) and al-Maġribī (11th/17th c.); for the former, see Raggetti, "*Cum grano salis*" (cit. note 9), pp. 334–337. For the relevant chapter in Ibn al-Ǧazarī (751–833/1350–1429), see Lucia Raggetti, "Inks as Instruments of Writing. Ibn al-Ǧazarī's Book on the Art of Penmanship," *Journal of Islamic Manuscripts*, 2019, 10:201–239.

16 A complete translation for the parts of the treatises related to ink making is in Fani, *Le arti del libro* (cit. note 1), pp. 39–154.

17 Muḥammad ibn Zakariyyāʾ al-Rāzī, "Zīnat al-kataba," edited by Luṭf Allāh al-Qārī, *ʿĀlam al-Maḫṭūṭāt wa-l-Nawādir*, 1432/2011, 16/2:211–242. See also Mahmoud Zaki, "Early Arabic Bookmaking Techniques as Described by al-Rāzī in His Recently Rediscovered *Zīnat al-Katabah*," *Journal of Islamic Manuscripts*, 2011, 2:223–234.

inks and invisible inks.[18] The work also includes recipes for treatments related to writing supports, such as the sizing of the paper and the reuse (as *ṭirs*, 'palimpsest') or antiquing of parchment, as well as for the removal of traces of ink or other substances from writing surfaces and fabrics. Indeed, the largest part of the work is dedicated to this latter subject. Other sporadic indications include hair dyeing and the sharpening of knives and swords. Looking at the unique witness of the work, the text is not visibly divided into chapters, but every recipe is introduced by an identifying heading.

The work *'Umdat al-kuttāb wa-'uddat ḏawī al-albāb* ('The staff of the scribes and the implement of the wise men'), has long been considered the most ancient treatise dedicated to the art of the book and, as such, has aroused much interest in both the Arab world and the West; this particular attention is underpinned by the existence of numerous witnesses to the text still preserved in libraries, dating back to different periods and coming from different geographical areas.[19] The treatise is traditionally ascribed to al-Mu'izz Ibn Bādīs (399–454/1008–1061), the Zirid governor of Ifrīqiya or to his son Tamīm (d. 501/1108), but it was probably compiled within the entourage of their court and dedicated to them. Three different recensions of the work have been identified by Armin Schopen on the basis of the additional chapters juxtaposed with the section on ink making.[20] Nonetheless, the number and the typology of recipes for ink making varies according to the different recensions; ferrogallic and mixed inks are largely favoured in the modern recensions, while the number of carbon inks is higher in the oldest one.[21]

The part of the *'Umdat al-kuttāb* related to ink production was explicitly mentioned as the basis of another treatise entitled *Al-muḫtara' fī funūn min al-ṣuna'* ('The findings on the techniques of craftsmanship'), by al-Malik al-Muẓaffar Yūsuf al-Ġassānī (m. 694/1294), the Rasulid sultan of Yemen. Four witnesses to *Al-muḫtara'* are known and an edition has been published.[22] The clearest difference between *Al-muḫtara'* and the Tunisian work is that, besides the chapters on the book arts (which, as anticipated, roughly reflects the *'Umdat al-kuttāb*), the *Al-muḫtara'* contains sections on martial arts, in particular

18 Al-Rāzī also included a recipe for an invisible ink in his *Kitāb al-ḫawāṣṣ*; see Raggetti, "*Cum Grano Salis*" (cit. note 9), p. 325.

19 The reference edition here is al-Mu'izz Ibn Bādīs, "'Umdat al-kuttāb wa-'uddat ḏawī al-albāb," edited by 'Abd al-Sattār al-Ḥalwaǧī and 'Alī 'Abd al-Muḥsin Zakī, *Maǧalla ma'had al-maḫṭūṭāt al-'arabiyya*, 1971, 17: 43–172. For the witnesses and other editions available, see Schopen, *Tinten und Tuschen* (cit. note 4), pp. 28–32.

20 *Ibid.*

21 Colini, *From recipes* (cit. note 3), p. 36.

22 Al-Malik al-Muẓaffar Yūsuf b. 'Umar b. 'Alī b. Rasūl, *Al-muḫtara' fī funūn min al-ṣuna'*, edited by Muḥammad 'Īsà Ṣāliḥiyya, (Kuwayt: Mu'assasat al-širā' al-'arabī, 1989).

on the construction of weapons such as the catapult and other battlefield furnishings. These sections are oddly juxtaposed with the art of book binding, in addition to chapters on fabric dyeing and stain removal, the production of soap and gold- and silver plating.

A less studied work is the treatise *Kitāb al-azhār fī ʿamal al-aḥbār* ('The most beautiful flowers on the production of inks'), by Muḥammad Ibn Maymūn al-Marrākušī al-Ḥimyarī, (mid-13th century). The only witness known to date is an autograph dated 649/1251–2 and compiled by the author during his stay in the *madrasa al-Mustanṣiriyya* in Baġdād.[23] Another autograph preserved in the Bibliotheque Nationale de France (MS Paris BnF Ar. 6915) attests to the author's competence in the field of alchemy. Unfortunately, of the 27 chapters listed in the introduction of the *Kitāb al-azhār*, only six are actually compiled, after which the author stops writing, consumed by an emotional crisis due to an unrequited love. The topics covered in his treatise are limited to the preparation of different types of liquid and dried black inks and coloured inks.[24]

The last work for the period to be considered is the *Tuḥaf al-ḫawāṣṣ fī ṭuraf al-ḫawāṣṣ* ('The gifts of the wise men on the curiosities of the substances') by Abū Bakr Muḥammad b. Muḥammad al-Qalalūsī (m. 607–707/1210–1307), an Andalusian scholar mentioned in different bibliographical repertoires; he was an authority in the field of Arabic language and a poet. In the introduction to his treatise, he dedicates it to a *kātib*, a chancellor who worked at the Nasrid court of Granada. There are currently two known witnesses to this work (one in Rabat, the other in the National Library in Paris) according to which an Arab edition was published in 2007.[25] The two witnesses present a different sub-division of the text: the one kept in Rabat is divided into three chapters

23 It has been published as a photostatic reproduction, introduced by Ibrāhīm Šabbūḥ: Muḥammad b. Maymūn b. ʿImrān al-Marrākušī, "Kitāb al-azhār fī ʿamal al-aḥbār li-Muḥammad b. Maymūn b. ʿImrān al-Marrākušī (al-qarn al-sābiʿ al-hiǧrī)," *Maǧallat taʾrīḫ al-ʿulūm al-ʿarabiyya wa-l-islāmiyya = Zeitschrift für Geschichte der Arabisch-Islamischen Wissenschaften*, 1422/2001, 14:41–133 (Arabic pagination).

24 According to the initial index, the missing essays would have been dedicated to: metallic inks, also those for writing on metal surfaces; the plating of various materials; the decoration of the hands with special dyes; the combination of the various pigments; the variations of writing colours over time; possible gilding surrogates; invisible inks; the removal of stains and erasing writing from writing supports; the method for breaking and restoring seals; different treatments of paper; the manufacture of the *līqa* (the felt pad on the lip of the inkwell); as well as a couple of sections dedicated to iron manufacture and to the annealing of swords and knives and a section dedicated to stain removal. There is also the title of a section dedicated to theoretical speculations about the rational scientific method. Al-Marrākušī, *Kitāb al-azhār* (cit. note 23), pp. 67–70.

25 Abū Bakr Muḥammad b. Muḥammad b. Idrīs al-Quḍāʿī al-Qalalūsī, *Tuḥaf al-ḫawāṣṣ fī ṭuraf al-ḫawāṣṣ*, edited by Ḥusām Aḥmad Muḫtār al-ʿAbbādī (al-Iskandariyya: Bibliotheca Alexandrina, 2007).

concerning, respectively: ink making; ink and stain removal from different surfaces; the use of ink on solid metals, dyes and paints. The Paris witness, by contrast, is divided into two chapters, the first concerning the preparation of inks, the second the removal of stains.

4 Approach to the Texts and to Their Transmission

The intellectual approach to this kind of textual production, equally well established in the Western literary tradition, was primarily based on the technical content that the sources are thought to deliver.[26] According to the approach of the 18th-century Italian erudition and then expanded throughout Europe, the technical sources were essentially considered a guide containing practical instructions that could be interpreted and reproduced through the identification of the chemical components of the materials employed. From the 19th c., the study of these testimonies of artistic practices aimed at the restoration of ancient artefacts with the intention to repair or integrate the original objects in a completely mimetic style. Today, the approach to restoration has, of course, changed radically and is now aiming at safeguarding the conservation of objects through the employment of reversible products and recognizable integration techniques without cancelling the stratifications and the traces of their material history. It is only in recent decades that the dissolution of the traditional functional interpretation of these sources, which anchored them to the technical information they preserve, opened up a wider, multidisciplinary approach that allows for appreciation of their wider cultural significance. The traditional assumption according to which textual sources related to technical crafts would have represented the most reliable proof of the practices and the materials employed by artists at the time these treatises were compiled, has been questioned on the basis of textual and philological considerations.[27] This open attitude towards such literary productions allows us to investigate the idea of what was intended to be transmitted, by whom and to whom, their knowledge and competences, and the cultural and social contexts within which these texts circulated.

26 Simona Rinaldi, "Per una filologia dei trattati e ricettari di colori," *Studi di Memofonte*, 2016, 16:1–16.

27 Sandro Baroni, Paola Travaglio, "Considerazioni e proposte per una metodologia di analisi dei ricettari di tecniche dell'arte e dell'artigianato. Note per una lettura e interpretazione," *Studi di Memofonte*, 2016, 16:25–83.

Such literary productions, namely technical treatises in the guise of compilations of recipes, are usually subject to fluid transmission, reflected in the presence of different recensions and branches of transmission of each work. The structure of the texts, arranged as a juxtaposition of different textual units relating to specific technical procedures fosters the production of variants at different levels — i.e. in the number and selection of the recipes, in the macrostructure of the treatise and the arrangement of the recipes, but also within single recipes — depending on the different skill levels of their copyists or compilers. An analysis of the available sources reveals that the textual units (recipes) or groups of units often transcended the boundaries of a specific textual tradition and were included in later compilations. In fact, in the classical Arabo-Islamic literary context, as in other pre-modern literary contexts, the concept of authorship was very blurred, and quotations from previous sources were common and not exclusive to treatises composed as recipe collections.[28] Thus, in addition to the essential and desirable critical editions of the works in question, a comparative textual analysis is crucial.[29] In a wider perspective, this comparison should not be limited to the treatises in the Arabic language, but should also include other historical and cultural contexts, as suggested by the evident echoes of late antique and early medieval sources found in the treatises under study.[30]

As mentioned above, the roles of author, compiler and copyist seem to intertwine and cross each other in the different phases of transmission of this technical material, contributing to very fluid and elastic traditions in which the number of recipes and their organization is not stable.[31] The identification of the authorial components of each work and their distinction from editorial or compilatory elements is crucial for a correct interpretation of these kinds of sources, not only to investigate the mechanisms of their formation and transmission, but also to evaluate the technical content elaborated and delivered by every author.

28 Sandro Baroni, "Ricettari: struttura del testo e retorica," *Studi di Memofonte*, 2016, 16:90–113, p. 90.

29 Sound critical editions of these fluid and unstable traditions are challenging to establish. For an exemplar critical edition of an Arabic compilation of recipes, see Raggetti, *ʿĪsā ibn ʿAlī's Book* (cit. note 4).

30 Parallels with the Western tradition can be found both in textual procedural elements as well as in the structure of the compilations themselves and in the different subjects juxtaposed with sections on colour/ink making. Fani, *Le arti del libro* (cit. note 1), pp. 175–180. Cf. Baroni, Travaglio, "Considerazioni" (cit. note 27), pp. 26–27.

31 *Ibid.*, pp. 30–31.

5 Identification of the Authorial Components

A fruitful preliminary consideration focused on the supposed competence of the alleged authors, informed by the available biographical data, by what they declared in the introductions to their works and, finally, by the actual technical content they transmit. The analysis thus includes both the macrostructure of the works (the treatises' 'narrative' organization) and their microstructures (i.e. the recipe units).

The high number of recipes transmitted by each work, and the juxtaposition of rather different subjects, such as ink making, bookbinding, paper making, but also stain removal, fabric dyeing, soap making and even the production of battlefield equipment and weapons, makes attribution to a single author at least questionable. The cases of the treatises ascribed to the Zirid ruler al-Muʿizz Ibn Bādīs and to the Rasulid al-Malik al-Muẓaffar are quite emblematic.[32] If, on the one hand, it is difficult to imagine that rulers could dominate such different sorts of crafts with such technical competence, and that they had the time to collect all these materials, then, on the other hand, their position would certainly have facilitated their access to written sources in the libraries of their palaces and their oral communication with experts. At the same time, the introductions to the two works attest explicitly to their compilatory nature. The ascription to the two rulers can be reasonably read as an act of deference by one of the *ʿālim* of their entourages. Moreover, the Rasulid ruler, or whoever compiled the treatises on his behalf, explicitly mentions Ibn Bādīs (referring to him as *ṣāḥib al-Mahdiyya*, 'the governor of al-Mahdiyya') as a source for the part of his treatise related to ink making.

The well-attested biographical and scientific profiles of al-Rāzī and al-Marrākušī certainly grant them a presumed higher degree of competence as alchemists and physicians, two fields that share with ink making common a knowledge related to the transformation of natural elements. Even in these cases, however, and in particular in relation to al-Marrākušī, the high number of recipes transmitted, sometimes very similar to each other, strongly suggests a possible, at least partial, compilatory nature of his work. At the same time, the *Kitāb al-azhār*, more than the others, appears to be a text closely related to a real operative context, as attested by the arrangement of the recipes according

32 It is worth mentioning that other works are ascribed to these two rulers, attesting to their possible competence in different fields of sciences, but also a possible common practice for the attribution of literary and technical works of interest to rulers; Fani, *Le arti del libro* (cit. note 1), pp. 155–164.

to the process used in the treatment of ingredients and by the numerous interventions of the author in terms of practical advice.[33]

The organization of the treatises and the categories in which the inks are included are very interesting both from the perspective of assessing the authors' competence and for understanding the modalities of transmission of these texts. The deconstruction of these texts at different levels, from the chapters and sub-chapters to single recipes, enhances our understanding of their origin and allows us to identify, at least in some cases, the sources from which they developed. At the same time, it reveals their specific peculiarities and the cultural context to which they refer. As for the classification of the inks, the basic structure of the recipe usually includes a short title, which may refer to different aspects of the product obtained or to its manufacture.[34] Nonetheless, it is worth underlining that rubricated titles were commonly added after the copy of the text, in blank spaces left by the scribes and in some cases filled in by someone else. This practice may have resulted in a lack of consistency between the title and the content of the recipe, with the titles changing or shifting from one recipe to another during the various stages of copying. For this reason, it is misleading to consider inconsistences between the title of a recipe and its content as a crucial proof of an authors' lack of competence on the subject. A general distinction is usually made between carbon inks (traditionally defined as *midād*), iron-gall inks (*ḥibr*) and mixed ink (*midād/ḥibr murakkab*), all encompassing both black and coloured inks.[35] An additional category common to the treatises is that of the coloured inks or paints (*līqa*), which were possibly employed by tapping the felt or silk pad or tuft of cotton on the lip of the inkwell, to which the term refers.[36] As previously mentioned, the degree of coherence between the definition given in the titles and the actual product described in the text could have been subject to changes made during the history of the text's transmission and is not necessarily an authorial responsibility. A meaningful example is represented by al-Qalalūsī's treatise in which the product of every recipe is designated in the title as *midād*, notwithstanding the actual ingredients and the ink obtained. Interestingly, al-Qalalūsī

33 Fani, *Le arti del libro* (cit. note 1), pp. 171–172.

34 See *infra*, p. 115–116.

35 Cf. Ibrāhīm Šabbūḥ, "Maṣdarāni ǧadīdāni ʿan ṣināʿat al-maḫṭūṭ: ḥawla funūn tarkīb al-midād" *Dirāsāt al-maḫṭūṭāt al-Islāmiyya bayna iʿtibārāt al-mādda wa-l-bašar: aʿmal al-muʾtamar al-ṯānī li-Muʾassasat al-Furqān li-l-Turāṯ al-Islāmī, 1314 / 1993*, edited by Rašīd al-ʿInānī (London: Muʾassasat al-Furqān li-l-Turāṯ al-Islāmī, 1417/1997):15–34, p. 23.

36 For a description of these categories, see Colini, *From Recipes* (cit. note 3), pp. 17–22. In the treatises, the term *līqa* indicates coloured inks without gall nuts (sometimes also called *midād*), for which a more common employment using the tuft or pad can be assumed.

also labels the recipes belonging to other traditions, such as those explicitly ascribed to al-Rāzī and those defined as *ḥibr* in the *Zīnat al-kataba,* as *midād*.[37]

As mentioned, in some cases, i.e. al-Rāzī's recipes included in al-Qalalūsī's work, it is the title of the recipe itself that allows us to trace the previous source. Only some of them are attested in the unique witness we have of the work *Zīnat al-kataba,* suggesting the idea that it could be an excerpt of a wider text. In general, we cannot automatically assume that the characters mentioned in the headings have produced literary works including ink recipes. It is also possible that these personalities transmitted the recipe orally, or that the ink described was associated to their names because they used to prepare it for their personal use — in this case, the description of the procedure could have been transmitted by someone else — or even that a pseudo-epigraphic attribution was made to an authoritative source. It is therefore misleading to think that every reference to famous personalities appearing in the titles of single recipes testifies to these people inventing or transmitting them in a written form without supplementary confirmation in the textual traditions.[38]

6 Redactional Interventions

If the authorial and original nature of the compilations can be supported by some of the mentioned elements related to the macrostructure of the treatises (i.e. authorial introductions to the texts and the systematic organization of the recipes according to different criteria), other textual elements at a micro-level of analysis could contribute to the identification of their compilatory additions and redactional extensions.[39] In this case, we consider an "authorial" or "original" compilation to be a set of recipes that have been described and transmitted by someone according to his personal thinking and with his own original writing, on the base of a direct observation or experience of the procedures, or after having obtained them via oral or written transmission.[40] This is testified at a formal and microstructural level by recipes with very similar syntactical structure, where the vocabulary is extremely reduced and repetitive. The

---

37 al-Qalalūsī, *Tuḥaf al-ḫawāṣṣ* (cit. note 25), pp. 21–22, 23–24; Fani, *Le arti del libro* (cit. note 1), pp. 172–173 (recipes Q 1.4, 5, 11). It is not possible to establish at what point during the transmission of this text the lexical change occurred, but it is possible that the change was made on the basis of a different use and perception of these technical terms in the peripheral region of al-Andalus.

38 Cf. Šabbūḥ, *Maṣdarāni ǧadīdāni* (cit. note 35), p. 22.

39 Baroni, Travaglio, *Considerazioni* (cit. note 27), p. 31.

40 *Ibid.*, p. 30.

authorship can refer to the entire compilation or to a certain core of recipes to which other material has been added drawing from other written sources.

6.1 *Syntactical Structure*

Regarding the lexical aspect of the treatise, a couple of examples show how the presence of dialectal terms can help in the identification of the authorial recipes. The first case is represented by al-Malik al-Muẓaffar's treatise. As mentioned above, the *ʿUmdat al-kuttāb* is the explicitly mentioned source of the Yemeni work for the part related to ink making. Of the 28 recipes for carbon inks in the Tunisian treatise, only six are included in *Al-muḫtaraʿ*, one of which proves to be original, at least in its textual dimension, and is characterized by the use of a Yemeni dialectal term to indicate sesame oil (*salīṭ*), not mentioned in other sources.[41] In the same way, the Andalusian al-Qalalūsī also introduces the dialectal name of a plant to obtain a red pigment: the *riğl al-ḥamām*, ('leg of the dove'), to indicate the *Anchusa tinctoria* L., the use of which is not mentioned in other treatises.[42]

Regarding the syntactical structure of the recipes, there are no apparent discrepancies among the different works. The most common scheme is scanty and characterized by a set of textual elements referring to the definition of the product, the name of the ingredients and their quantity, the different procedural phases with additional specification about the modalities, duration, place of the operations etc., and finally the result or the indication of the use of the product, as shown by the example in the following chart:[43]

Title (MM II.18)	صفة حبر وردي	*Description of a pink ink*
Starting verb I ("to take")	تؤخذ	*It is taken*
Specification of quantity I	اوقية	*an* ūqiyya
Ingredient I	سيلقون	*of minium*
Operation I (verb)	فيسحق	*and is crushed*
Specification of place	على بلاطة	*on a slab of stone*
Operation II (verb)	ويلقى	*and it is added*
Specification of quantity (II)	وزن درهم	*the weight of a* dirham

41 In fact, the recipe transmitted is a carbon ink obtained from the soot of sesame oil, which was quite common in other Arab treatises and is described as an Egyptian ink. Al-Malik al-Muẓaffar, *Al-muḫtaraʿ* (cit. note 22), p. 70; Fani, *Le arti del libro* (cit. note 1), pp. 59, 169 (recipe MM II.6).

42 al-Qalalūsī, *Tuḥaf al-ḫawāṣṣ* (cit. note 25), p. 28; Fani, *Le arti del libro* (cit. note 1), pp. 146, 189 (recipe Q II.4).

43 Al-Malik al-Muẓaffar, *Al-muḫtaraʿ* (cit. note 22), p. 76.

Ingredient II	بورق	*of borax*
Specification of quantity III	ودرهمين	*and two* dirhams
Ingredient III	صمغ	*of gum arabic*
Operation III (verb)	ويدلك	*and it is kneaded*
Specification of mode	حتى ينعم وهو يسقى	*until it becomes fine while it is soaked*
Ingredient IV	بماء العفص	*with a solution of gall-nut*
Result and use	ويكتب به	*and it is written with it*

6.2 *Lexical Elements*

Apart from the few examples of dialectal terms employed, the vocabulary of the recipes is quite repetitive across the different traditions; this is true not only for the names of the ingredients, but also for the procedural verbs used in the sources. They can appear both in the jussive form and in an impersonal form (expressed by the use of a passive form); the alternation is quite frequent not only within a single work, but also within the same recipe, thus the shift does not seem to be related to the interpolation of recipes from different textual traditions. Moreover, operations aimed at a more formal homogeneity of the material gathered in a compilation of recipes can also be undertaken by the compilers or copyists. If this can be considered a rather passive intervention, affecting only the formal dimension of the recipes, in other cases the scanty and telegraphic structure of a recipe can also be affected by substantial redactional interventions or rewriting, which expand and/or transform the original formulation of the technical procedure. Editorial interventions of this kind can occur not only within a single textual tradition in the process of copying the work, but also in the transmission of textual units from one tradition to another, so that the same recipe, or better, the common origin of different recipes, sometimes becomes barely recognisable.[44] The example shown in the Appendix is meaningful in two respects: one the one hand, it shows that the later authors should know al-Rāzī's treatises; on the other hand, it gives a sound idea of the competences of the person who intervened in the original recipe.[45]

6.3 *Organization of the Compilations*

In addition to the modifications to the text of single recipes, the works emerging from an editorial or redactional process can show interventions in the

44 See Appendix.

45 Sometimes, additional phrases attesting to the utility or the effectiveness of a procedure represent mere literary *topoi*; take, for example, the word *muğarrab*. Cf. Raggetti, *ʿĪsā ibn ʿAlī's Book* (cit. note 4), pp. xv–xvi.

macrostructure of the collection, when the order of the recipes is adapted to the editor's different *consecutio*. The classifications and categories of inks attested in the sources available are expressed alternatively in the title of the recipes or in the headings of chapters or sub-chapters in which the recipes are included, and referred to:

- the colour of the products obtained (black, different colours, metallic, invisible);
- their composition (*ḥibr*/iron-gall inks, *midād*/carbon inks, *murakkab*/ mixed);
- their final state (liquid, solid, powder);
- their preparation (under the sun, on the fire, in the shadow, by infusion, by decoction, by maceration)
- their final employment (with reference to the writing support, to the texts to which their use is dedicated, to the application implement, for example the *līqa*);
- the famous historical characters and scholars who used / transmitted the recipe.

Theoretically, any recipe can be integrated in one or more of these categories simultaneously; thus, it is common for an author who extrapolates recipes from a previous source to insert it into the appropriate category of his own work. In the *Kitāb al-azhār*, for example, al-Marrākušī had to adjust the order of the recipes derived from al-Rāzī's and Ibn Bādīs's works so that it was compatible with his highly original scheme arranged according to the method of preparation and manipulation of the ingredients.[46]

While in certain cases the inclusion of recipes from previous sources is made explicit in the text itself — for example in *Al-muḫtaraʿ fī funūn* —[47] or in the titles of the recipes, in other cases, only comparative analysis can clarify the phenomenon and identify the hypothetical original/authorial version of textual units or, better put, the first occurrence in written sources of this specific linguistic context.[48]

46 Cf. *supra*, note 24.

47 Cf. *supra*, p. 110, 114.

48 The recipes can be considered original only on the basis of the available sources, but they are certainly a much smaller number of what has been written. In any case, possible syntactical or lexical discontinuity in what is considered to be the original *nucleus* of a work, or possible historical or regional elements attesting a more ancient origin of a recipe within that core, or a geographic origin different from that of its alleged author, should hint at a derivation from a different textual tradition.

6.4 *Explanatory Additions*

Another common redactional intervention in the treatises studied is the addition of errant phrases or explanations of difficult passages in recipes derived from different traditions; a meaningful example is the note added by al-Marrākušī to the margin of his treatise, in correspondence of one of the recipes ascribed to al-Rāzī. The note refers to the word *istār*, a unit of measure not previously mentioned in other treatises and that evidently required the equivalence to the system more commonly used in Baġdād.[49] In another case, the same al-Marrākušī made a redactional intervention in the transmission of a recipe ascribed to Ibn Buḫtīšūʿ and also mentioned in al-Rāzī and in al-Qalalūsī treatises, in the latter with an explicit attribution to the same al-Rāzī.[50] In this case, al-Marrākušī changes the unit of weight measurement attested in the other two treatises, i.e. *mikyāl*, replacing it with a proportional system of parts. This was probably done to avoid confusion with the mention of an uncommon reference system.[51]

6.5 *Modalities of Formation*

More broadly, these phenomena can give us an idea of the modalities of formation of these kinds of collections. In fact, the texts can be extended in various aggregative ways, or, on the contrary, can be reduced, contracted or deconstructed and rearranged. And, even if, generally, in this kind of production "sobriety is almost always primitive,"[52] a comparison of the texts reveal a more elastic dynamic in the history of their transmission. Just to make some examples, the 11th-century treatise *ʿUmdat al-kuttāb*, appears to be the richest for number of recipes transmitted (approximately 150), despite its antiquity. At the same time, if we focus on the text of single recipes, like the one presented in the Appendix, the 13th-century treatise *Kitāb al-azhār* presents an expanded version of the material derived from previous sources. Nevertheless, its almost

---

49 The note reads: "al-Marrākušī says, God bless him: there are 40 *istār* in the *mann* of Baġdād."; al-Marrākušī, *Kitāb al-azhār* (cit. note 23), p. 75.

50 *Ibid.*, p. 80; al-Rāzī, *Zīnat al-kataba* (cit. note 17), pp. 225–226; al-Qalalūsī, *Tuḥaf al-ḫawāṣṣ* (cit. note 25), p. 21. The reference is to the famous family of Persian physicians who worked between the 7th and the 9th centuries; they all came from the Gundishapur Academy and some of them became personal physicians to the Abbasid caliphs; see Dominique Sourdel, "Bukhtīshūʿ," in *Encyclopaedia of Islam, New Edition* (cit. note 5), vol. I p. 1298. Armin Schopen suggests the identification with Ǧibrīl Ibn Buḫtīšūʿ, see Schopen, *Tinten und Tuschen* (cit. note 4), p. 89.

51 For an excursus on the units of measure used in the treatises under study, see Fani, *Le arti del libro* (cit. note 1), pp. 186–187.

52 Baroni, Travaglio, *Considerazioni* (cit. note 27), p. 33.

contemporary treatise, the Yemeni *Al-muḫtaraʿ fī funūn min al-ṣunʿa* and the Andalusian *Tuḥaf al-ḫawāṣṣ fī ṭuraf al-ḫawāṣṣ*, presents a much scantier version of the same units.

The modalities of formation of this kind have been identified following the study of Western treatises on colour making and can easily be adapted to treatises in the Arabic language.[53] Among the modalities of expansion of texts, the most common process of aggregation in our treatises is the interpolation of selected recipes from different sources based on the different categories of inks elaborated by every author. This aggregative system is also reflected in the titles of the recipes and they are often introduced using the heading *ṣifa uḫrà* (another description), or similar phrases, following a more precise definition of the first recipe of a set. Only in the case of invisible inks is it possible to recognize a thematic aggregation in almost all the sources at our disposal.[54] The set of recipes "to put secrets in the books" (*waḍaʿ al-asrār fī al-kutub*), as they are generally described by the sources, is already attested in the more ancient treatise, the *Zīnat al-kataba* by al-Rāzī, and it is mentioned in all the subsequent compilations, even if the recipes do not appear exactly in the same order within the set.

7 Literary and Rhetorical Elements

A final consideration will focus on the introductions to these works. If, on the one hand, they frequently refer to the circumstances of their composition, revealing important clues about the real technical competence of the authors/compilers, they provide evidence of a literary dimension and of their intellectual framework. The presence of rhetorical elements in the introductions has been underlined and categorized based on previously conducted studies of specific Western treatises.[55] The same categories and rhetorical elements typical of literary works can be identified in the introductions to the Arabic treatises, demonstrating a common cultural philosophical background. The specific constitutive elements have been reapplied by Sandro Baroni to the

53 *Ibid.*, pp. 33–52.

54 In the *Kitāb al-azhār*, the chapter on invisible inks is not developed, despite being announced by the author in the initial index; cf. *supra*, note 24.

55 Maite Rossi, "Il pensiero e il colore. Modelli della filosofia classica nella letteratura tecnico-artistica medievale," *Quaderni dell'Abbazia. Fondazione Abbatia Sancte Marie di Morimundo e Museo dell'Abbazia di Morimondo*, 2008, 15:161–192.

common rhetoric scheme of the Aristotelian *exordium*.[56] This can be summarized as:

- *Prótasi*: self-presentation — declaration of intent / recipient / subject;
- *Invocatio*: invocations to the divinity — theological references — invocation to the dedicatee;
- *Captatio*: statements of humility — reference to an authority — declaration of simplicity / truthfulness / experience.

Below is a sample of the introduction from the *Zīnat al-kataba* that highlights the constituent rhetorical elements:[57]

Invocation to the divinity	In the name of God, the most Gracious, the most Merciful
Subject	This is the book, the adornment of the scribes and what they cannot live without.
Declaration of intent and declaration of utility	It can happen that [the content of] one of its chapters is needed [by someone] with no proficiency in the subject, so the resulting damage is big and the *kātib* is discredited.
Theological reference	We have therefore put together what this book consists of, "and above every person of knowledge, there is one more learned" [Qur. XII: 76].

The same elements can be identified in the longer introduction of al-Marrākušī's work.[58] In this regard, Baroni underlines that Aristotle supports an indirect proportionality between the elaboration and length of this part of the speech and the technical competence of the audience, stating that a *proemio* or *exordium* introducing the subject is not necessary for a competent public.[59]

8 Conclusions

To conclude, according to a multi-level analysis, exemplified here by specific case studies, the sources under investigation appear to provide much more information than the procedural indication for ink production. As we

56 Baroni, *Ricettari* (cit. note 28), p. 104.

57 Cf. the Arabic text in al-Rāzī, *Zīnat al-kataba* (cit. note 17), p. 222.

58 Cf. Fani, *Le arti del libro* (cit. note 1), pp. 80–81, 157–160.

59 Baroni, *Ricettari* (cit. note 28), pp. 104–10; see also, Roland Barthes, *La retorica antica* (Milano: Bompiani, 1972), pp. 91–92.

have tried to demonstrate, their technical content as well as their structure and organization at a micro- and macro-level, clearly refer to well-established traditions and theoretical paradigms, which must be recognised in order to interpret the written expression of this professional knowledge. As Sandro Baroni writes:[60]

> The technical literature related to the arts is certainly the reflection of a practical knowledge in constant updating, but we must not forget that the literary texts that transmit this knowledge present all the problems originated by the distance between the word and the practice. They are submitted to expository and organizational methods of the "technical report," which are sometimes able to alter and distort the mere operative procedures, but at the same time contain further information about the generative context of the technical contents.

In other words, the sources presented here reflect the social history of the communities within which they were produced. Their organization, their technical content and the literary *topoi* are crucial elements requiring specific attention and interpretation in a comparative perspective.

Within the Arabic literary tradition, these treatises are located in the broad category of *adab* literature, in which the didactical or technical content of a text not only aims to instruct, but also to educate, in a wider sense, the literate class of the *kuttāb* and to show the knowledge available in that particular field.[61] The fact that, in some cases, these treatises could have been used as a guide for ink making by scribes or craftsmen in ink makers' workshops is plausible and undeniable, but the intention that animated these compilations certainly goes beyond the didactical and technical purpose of manuals or handbooks.

Appendix

R Abū Bakr Muḥammad b. Zakariyyāʾ al-Rāzī (d. 313/925), *Zīnat al-kataba*

IB al-Muʿizz Ibn Bādīs al-Ṣanḥāǧī (d. 454/1061), *ʿUmdat al-kuttāb wa-ʿuddat ḏawī al-albāb*

MM al-Malik al-Muẓaffar Yūsuf al-Ġassānī (d. 694/1294), *Al-muḫtaraʿ fī funūn min al-ṣunaʿ*

60 *Ibid.*, p. 106 (the English translation is mine).

61 Fani, *Le arti del libro* (cit. note 1), pp. 191–193.

MḤ Muḥammad b. Maymūn b. ʿImrān al-Marrākušī (fl. 649/1251–2), *Kitāb al-azhār fī ʿamal al-aḥbār*

Q Abū Bakr Muḥammad al-Qalalūsī (d. 707/1307), *Tuḥaf al-ḫawāṣṣ fī ṭuraf al-ḫawāṣṣ*

	R[a]	IB = MM[b]	MḤ[c]	Q[d]
١	المداد لا يمحى ولا يذهب اثره تكتب ان شئت لا يمحى من القرطاس اوكاغد او غير ذلك وهو خضاب للشعر ايضا	صفة مداد عراقي	هذه النسخة لاهل المغرب لا تمتحي من الرق ولا يذهب اثرها من الورق وتجدها مع ذالك خضابا للشعر جليلا عجيبا يبقي زمانا طويلا فافهمها منا يا أخي حَيّاكَ الله وَأبقاك	مداد لا ينقلع أبدا وهو أيضا خضاب ذكره الرازي من شقائق النعمان
٢	تأخذ شقائق النعمان	تؤخذ الشقائق	انك تَاخُذُ ما تريد من شقايق النعمان الشديد الحمرة على ما تقدم ذكره واسمه ان فهمت مكانه	تأخذ ما أحبيت
٣	وتحشوه قارورة رقيقة شامية	فتحشى في القوارير الدقاق	فتحشوه في قارورة رقيقة شامية أو مغربية أيهما كان	وتحشوه في قارورة رقيقة

a Al-Rāzī, *Zīnat al-kataba* (cit. note 17), pp. 222–223.

b Ibn Bādīs, *ʿUmdat al-kuttāb* (cit. note 19), pp. 82–83; al-Malik al-Muẓaffar, *Al-muḫtaraʿ* (cit. note 22), pp. 68–69.

c Al-Marrākušī, *Kitāb al-azhār* (cit. note 23), pp. 106–107.

d Al-Qalalūsī, *Tuḥaf al-ḫawāṣṣ* (cit. note 25), p. 25.

(*cont.*)

	R	IB = MM	MḤ	Q
٤	ثم تدفنه في سرقين رطب	وتدفن في سرجين	ثم تدفنه في سرجين وله من الاسماء والمعنا الدواب واحد في كتبنا المتقدّم ذكرها الموسومة بالحكمة السرقين والسرفين وبطون الحيل والطبيعة تخدم الطبيعة والنار الحايلة فافهمها من المرّاكشي هنا وهناك ويكون الزبل الرطب	ثم تدفنه في السرفين
٥	وتبدل سرقين كل ثلاث ايام		وتبدّل له الزبل الطري كل ثلاثة أيام وأنت تعتبره بعد ذلك	وتبدل السرفين كل ثلاثة أيام
٦	حتى تنظر اليه وقد ذاب وصار ماء وانحل	حتى تذوب وتصير ماء وتحل	حتى تراه قد ذاب وصار ماء منحلّا	حتى تنظر إليه وقد ذاب وصار ماء
٧	واكتب به حيث شئت	–	وهينئذ تكتب به حيث شيت فيما ذكرنا	ثم اكتب به حيث شيت
٨	فلا يمحى	–	فإنه لا يمتحى جملة واحدة	فإنه لا يمحي
٩	وان وقع في الماء اياما	–	ولن تقيى القرطاس المكتوب به في الماء أياما على ما ذكر والسلام	وإن دفع في الماء أياما

(*cont.*)

	R	IB = MM	MḤ	Q
١٠	–	–	قال المرّاكشي وهو محمد ابن ميمون بن عمران المرّاكشي الحميري سنة تسع وأربعين وستماية ببغداد والزيادة في هذا الباب	–
١١	وإن شئت أن يكون براقا فاجعل فيه صمغا عربيا	–	إن شيت أن يكون براقا فتجعل فيه من الصمغ العربي المسحون مقدارا مناسبا	وإن شيت أن يكون براقا فاجعل فيه صمغا
٢١	–	–	فإنه يجي مشرقا مليحا مونقا إن شاء الله تعالى	فإنه عجيب
			[...]	
٣١	–	ثم تعمد إلى القراطيس فتحرقها	–	–
٤١	–	وتجمع ما احترق منها بذلك الماء	–	–
٥١	–	وترفعه إلى أن يجف في الظل	–	–
٦١	–	ثم يؤخذ منه وزن درهم	–	–
٧١	–	ومن ماء الصمغ العربي وزن درهم	–	–

(*cont.*)

	R	IB = MM	MḤ	Q
٨١	–	ومن العفص المسحوق وزن نصف درهم	–	–
٩١	–	فيسحق الجميع ببياض البيض	–	–
٠٢	–	ويبندق ويجفف كما ذكرنا آنفا	–	–
١٢	–	وتحشى به الدواة عند الحاجة إليه مع ماء السلق وهو أجود ماء لها	–	–

	R	IB = MM	MḤ	Q
1	An ink that doesn't fade and whose traces don't disappear; if you want, write, and it will not fade on the papyrus, or paper, or other material; it is also a dye for the hair.	Description of an Iraqi ink.	This copy is of the people of Maghreb; it doesn't fade from the parchment and its traces don't disappear from the paper. You will discover that it is a beautiful and marvellous dye for the hair that lasts for a long time. Learn it from me, my brother, may God preserve you.	An ink that never fades and is also a dye; al-Rāzī mentioned it [as made] with anemone.

(*cont.*)

	R	IB = MM	MḤ	Q
2	Take some anemones	Take the anemones	Take the amount you want of dark red anemone as it was mentioned and named above, so that you can understand right away	Take the amount you like
3	and put them in a thin Syrian bottle	and put them in thin bottles	put it in a thin Syrian or Maghribi bottle, whichever it is	and put it in a thin bottle
4	then bury it in the humid dung	you bury them in the liquid dung	then bury it in the dung that has different names in my books mentioned above, but the meaning is one; it is wisely called *sirqīn* or *sirfīn*, or "the bowels of the horse," or "the nature useful to the nature," or "the unstable fire"; learn them all from al-Marrākušī [who wrote them] here and there; the manure has to be humid	then bury it in the dung
5	substitute the dung every three days	–	and you have to substitute it with other wet manure every three days	that you have to substitute every three days

(*cont.*)

	R	IB = MM	MḤ	Q
6	until you look at it and [you will see that] it has melted and become liquid	until they melt, become like water and liquefy.	then check it until you will see that it has completely melted and become liquid	until you look at it and [you see that] it has melted and has become liquid
7	then you can write with it wherever you want	–	then write with it wherever you want among the materials we have mentioned	then write with it wherever you want
8	and it will not fade	–	not even a sentence will fade	and it will never fade
9	not even if you put it the water	–	and, according to what is reported, the papyrus written with it will not be cleared out [if put] in the water for days	not even if you put it in the water for days
10	–	–	al-Marrākušī said – and he was Muḥammad b. Maymūn b. ʿImrān al-Marrākušī al-Ḥimyari – in the year 649 while he was in Baghdad, as an addition to this chapter:	–
11	if you want to make it bright, put in it gum arabic	–	if you want to make it bright, put the proper amount of powdered gum arabic in it	if you want to make it bright put some gum [arabic] in it
12	–	–	and it will be brilliant, beautiful and shiny [...]	and it will be marvellous
13	–	Then deal with the papyrus and burn it	–	–

(*cont.*)

	R	IB = MM	MḤ	Q
14	–	then mix its burned rests with that solution	–	–
15	–	and put it in the shadow to dry	–	–
16	–	take the weight of one *dirham* of it	–	–
17	–	one *dirham* of gum arabic solution	–	–
18	–	and half *dirham* of ground gall nuts	–	–
19	–	mash it with albumen	–	–
20	–	and obtain small balls from it and put them to dry as we have said above	–	–
21	–	when you need it, you fill the inkwell with this nuts with extract of chard, that is the best solvent for it.	–	–

CHAPTER 7

"I tried it and it is really good": Replicating Recipes of Arabic Black Inks

Claudia Colini

Abstract

The aim of this contribution is to present some case studies that highlight the key role played by the replication of recipes in both the Humanities and the Natural Sciences. In particular, I will present instances of how this approach can assist textual criticism, for example in the understanding of variants and errors, and in clarifying the meaning of some terms. By evaluating the feasibility of the recipes, their outcome and their order in the treatises, it is also possible to determine the technical skills of authors and compilers. Finally, the inks produced can be used as a reference for scientific analysis: not only by comparing these data with those obtained from the investigation of manuscripts, but also to assess the limits of the equipment, techniques and protocols used to undertake this investigation.

Keywords

Arabic manuscripts – ink recipes – scientific analysis

1 Introduction

The study of manuscripts' constituent materials has developed considerably in recent years. But alongside the use of increasingly refined non-destructive techniques and portable equipment, the need for an historiographical study of the materials known and employed for writing in the Arabo-Islamic milieu has become evident. In this view, historical sources, such as treatises on painting or calligraphy, can add information critical to the understanding of past epochs' technical knowledge and to the correct identification of the materials employed. But the study of these treatises must not be limited to their textual analyses and should instead be accompanied by the experimental reproduction of the recipes in order to yield the best results. Indeed, the replication of recipes can enhance understanding of the texts, which are sometimes unclear, and in evaluating the writers' competence in light of the corrections they made to the text they copied and of the different kinds of mistakes they might have introduced. Moreover, the inks produced can be used as a reference for scientific analyses: by knowing as precisely as possible which ingredients were used in a given recipe, it is possible to evaluate the effectiveness of protocols, techniques and instruments when attempting the identification of that ink's components. Furthermore, the recipes and the resulting inks can help explain abnormal or uncommon results observed through the application of scientific analysis on manuscripts.

With this interdisciplinary approach in mind, recent research has been conducted by collecting black ink recipes from written Arabic sources on bookmaking, then by assessing their feasibility, and finally by replicating some of them. These samples were artificially aged and analysed using an array of analytical techniques, mostly non-invasive and non-destructive, in order to build a database of Arabic inks and their ingredients and also to verify the limits of the portable equipment and of the techniques employed.[1]

The aim of this paper is to present some case studies that highlight the key role played by the replication of recipes in relation to a multitude of purposes: to spot possible mistakes in the tradition and to understand their transmission (case study I and II); to assess the author's technical skills (case study II, III and IV); to clarify the terminology (case study III and IV); to validate hypotheses based on analytical data (case study V); and finally, to assess the limits of the current analytical protocols and equipment (case study VI).

The main treatises taken into account in this chapter are:[2]

---

1 Claudia Colini, From Recipes to Material Analysis: The Arabic tradition of Black Inks and Paper Coatings (9th–20th century), (PhD Diss., Hamburg University, 2018). The research presented in this article was carried out at the SFB 950 'Manuskriptkulturen in Asien, Afrika und Europa,' funded by the German Research Foundation (Deutsche Forschungsgemeinschaft, DFG) and within the scope of the Centre for the Study of Manuscript Cultures (CSMC). I would like to acknowledge the help received by Sara Fani, Cornelius Berthold, Beate Wießmüller and Tilman Seidensticker for the translation of the Arabic texts. My thanks go to Ira Rabin, Tea Ghigo and Sebastian Bosch for the fruitful discussions and other members of our department at the BAM and CSMC for sharing their analytical results with me: Oliver Hahn, Olivier Bonnerot, Zina Cohen and Ivan Shevchuck. I would also like to thank Ahmed Parkar for his assistance in the replication of the recipes from Kenya.

2 The literary sources for this study are selected from the Italian translation by Sara Fani and in the German translation provided by Armin Schopen, as well as in their Arabic original. See Sara Fani, Le arti del libro secondo le fonti arabe originali. I ricettari arabi per la fabbricazione degli inchiostri (sec. IX–XIII): loro importanza per una corretta valutazione

- *Zīnat al-kataba* ('The ornament of the scribes') by Abū Bakr Muḥammad b. Zakariyyā' al-Rāzī, (d. 313 or 323/925 or 935);[3]
- *'Umdat al-kuttāb wa-'uddat ḏawī al-albāb* ('The staff of the scribes and implements of the wise men') by al-Mu'izz b. Bādīs at-Tamīmī al-Ṣanhāǧī (d. 454/1062);[4]
- *al-Muḫtara' fī funūn min al-ṣuna'* ('The findings on the techniques of craftsmanship') by al-Malik al-Muẓaffar Šams al-Dīn Yūsuf b. 'Umar al-Ġassānī (d. 694/1294–95);[5]
- *Kitāb al-azhār fī 'amal al-aḥbār* ('The most beautiful flowers on the production of inks') by Muḥammad b. Maymūn b. 'Imrān al-Marrākušī al-Ḥimyarī (7th/13th c.);[6]

---

e conservazione del patrimonio manoscritto, (PhD Diss., Università "L'Orientale," Napoli 2013); Armin Schopen, *Tinten und Tuschen des arabisch-islamischen Mittelalters*, (Göttingen: Vandenhoeck & Ruprecht, 2014); other recipes have been collected in the following precious contributions: Martin Levey, *Mediaeval Arabic Bookmaking and Its Relation to Early Chemistry and Pharmacology* (Philadelphia, PA: American Philosophical Society, 1962); Lucia Raggetti "*Cum grano salis*: Some Arabic Inks Recipes in their Historical and Literary Context," *Journal of Islamic Manuscripts*, 2016, 7/3:294–338; David James, "The Commentaries of Ibn al-Baṣīṣ and Ibn al-Waḥīd on Ibn al-Bawwāb *Ode on the Art of Calligraphy* (*rā'iyyah fī al-ḫaṭṭ*)," in *Back to the Sources. Biblical and Near Eastern Studies in Honour of Dermot Ryan*, edited by Kevin J. Cathcart & Jeremiah F. Healey, (Dublin: Glendale 1989), pp. 164–191; Hossam Mujtar al-'Abbādī, *Las artes del libro en al-Andalus y el Magreb (siglos IVh/XdC–VIIIh/XVdC)*, (Madrid: Ediciones El Viso, 2005); Michaelle Biddle, "Inks in the Islamic Manuscripts of Northern Nigeria: Old Recipes, Modern Analysis and Medicine," *Journal of Islamic Manuscript*, 2011, 2/1:1–35.

3 Lenn E. Goodman, "al-Rāzī," in *Encyclopaedia of Islam, New Edition*, edited by: P. Bearman, Th. Bianquis, C.E. Bosworth, E. van Donzel, W.P. Heinrichs, vol. VIII pp. 474a–477b; Carl Brockelmann, *Geschichte der Arabischen Litteratur*, Band I (Weimar: Emil Felber, 1898), pp. 233–235; Carl Brockelmann, *Geschichte der Arabischen Litteratur. Supplementband*, Band I, (Leiden: Brill, 1937), pp. 417–421; Mahmoud Zaki, "Early Arabic Bookmaking Techniques as Described by al-Rāzī in His Recently Discovered Zīnat al-Katabah," *Journal of Islamic Manuscripts*, 2011, 2:223–234.

4 Mohamed Talbi, "al-Mu'izz b. Bādīs," in: *Encyclopaedia of Islam, New Edition* (cit. note 3), vol. VII pp. 481b–484a; Brockelmann, *Geschichte* I (cit. note 2), p. 268; Brockelmann, *Geschichte Supplementband* I (cit. note 3), p. 473; al-Mu'izz ibn Bādīs al-Tamīmī al-Ṣanhāǧī, *'Umdat al-kuttāb wa-'uddat ḏawī al-albāb. Fīhi ṣifat al-ḫaṭṭ wa-l-aqlām wa-l-midād wa-l-liyaq wa-l-ḥibr wa-l-asbāġ wa-ālat al-taǧlīd*, eds Naǧīb Mā'il al-Harawī and 'Iṣām Makkiyya. Mašhad: Maǧma' al-Buḥūṯ al-Islāmiyya, 1409/1988 H.Sh.

5 G.R. Smith, "Rasūlids," in: *Encyclopaedia of Islam, New Edition* (cit. note 3), vol. VIII pp. 455a–457b.

6 Muḥammad b. Maymūn b. 'Imrān al-Marrākušī (al-Ḥimyarī), "Kitāb al-Azhār fī 'amal al-aḥbār li-Muḥammad b. Maymūn b. 'Imrān al-Marrākušī (al-qarn al-sābi' al-hiǧrī)," edited by Ibrāhīm Šabbūḥ, in *Maǧallat ta'rīḫ al-'ulūm al-'arabiyya wa-l-islāmiyya = Zeitschrif für*

- *Tuḥaf al-ḫawāṣṣ fī turaf al-ḫawāṣṣ* ('The gifts of the wise men on the curiosities of the substances') by Abū Bakr Muḥammad b. Muḥammad al-Qalalūsī al-Andalusī (d. 707/1308).[7]

2 Case Study I — A Recipe of a Plant Ink or of an Iron Gall Ink with a Missing Line?

In the first case study, I analyse one of the few recipes of plant inks recorded in the treatises. It has been found only in the treatise *Tanwīr al-ġayāhib fī aḥkām ḏawāt al-ḏawā'ib* ('The illumination of darkness concerning the rules of those [preparations] containing soluble substances'), attributed to al-Qalalūsī.[8] In my opinion, however, it is more likely that it is a combination of recipes from different sources, with the majority of recipes originating from *Tuḥaf al-ḫawāṣṣ* by al-Qalalūsī. In fact, while al-Qalalūsī called even the iron gall inks *midād* (a word used to identify mainly carbon inks), the term *ḥibr* is also used in *Tanwīr al-ġayāhib*, albeit not systematically. This is particularly the case for those recipes not coming from *Tuḥaf al-ḫawāṣṣ*.

> *The Ink* (ḥibr) *of Abū Ṭāhir*[9]
> It takes two and a half *dirham* of good gall nuts and the same amount of good gum. They are pulverized in a mortar. Twenty-five *dirham* of water are poured on it and it is rubbed until it is good and pure, God willing.[10]

---

geschichte der Arabisch-Islamischen Wissenschafen, 1422/2001, 14:41–133 (Arabic pagination), pp. 41–54.

7 Carl Brockelmann, *Geschichte der Arabischen Litteratur*, Band II (Weimar: Emil Felber, 1902), p. 336 (although his name is given as al-Qallūsī); Abū Bakr Muḥammad ibn Muḥammad al-Qalalūsī al-Andalusī, *Tuḥaf Al-Jawāṣṣ Fī Turaf Al-Jawāṣṣ* (*Las galanduras de la nobleza en lo tocante a los conocimientos más delicados*), ed. Hossam Ahmed Mokhtar El-Abbady. Alexandria: Maktabat al-Iskandariyya, 2007.

8 Schopen, *Tinten und Tuschen* (cit. note 2), p. 27; MS Cairo Dār al-Kutub *'Ulūm ma'āšiyya* 46.

9 Schopen identified the person as Abū al-Faḍl Aḥmad b. Abū Ṭāhir Ṭayfūr (204/819–280/893), writer, scholar and historian; Schopen, *Tinten und Tuschen* (cit. note 2), p. 125, Brockelmann, *Geschichte* I (cit. note 2), p. 144; Brockelmann, *Geschichte Supplement Band* I (cit. note 3), p. 210; Fuat Sezgin, *Geschichte des arabischen Schrifttums*, Band I, (Leiden: E.J. Brill, 1996) p. 348.

10 See Schopen, *Tinten und Tuschen* (cit. note 2), p. 125; for the Arabic text, see MS Cairo Dār al-Kutub *'Ulūm ma'āšiyya* 46, f. 31r.

When I replicated this recipe, the colour of the ink was pale yellow and the writing on the paper almost invisible. According to Schopen, a similar recipe (that he does not translate) was used to write on parchment under the Zirid rule in Qayrawān (Tunisia).[11] He himself tested Abū Ṭāhir's recipe on parchment obtaining a more visible ink, although it is still of a light yellow-brown colour.

Let us consider the steps of the recipe: 1) gall nuts and gum are pulverised in a mortar; 2) water is added; and 3) after rubbing and pounding the liquid for an unspecified length of time, the ink is ready. The solubilisation of tannins into water, however, is not immediate, something that the authors of ink recipes knew very well. In fact, in the recipes for iron gall inks, the tannins are often extracted by maceration, fermentation or cooking. In a few cases, the plants are used in the form of powders: these are the so-called instant ink (*ḥibr min sāʿatihī*) or shadow ink (*ḥibr ẓillī*).

As an example, I present here two recipes of this kind, both found in the same *Tanwīr al-ġayāhib* treatise as the aforementioned one by Abū Ṭāhir, respectively two folios before and two folios after it:

> *Good Ink* (midād) *for Paper*
> Two dirhams of Byzantine gall nuts, the same quantity of gum arabic and one and a half dirham [of vitriol] must be taken and ground separately, then, twenty dirhams of water must be added to the gum after having ground it in a mortar, and it must be left until it melts. Then the ground gall nuts must be added to it, and then the vitriol, God willing. For parchment: the preparation is the same, with the addition of four dirhams of water and the subtraction of one dirham of gum.[12]
>
> *Recipe for an Ink* (ḥibr) *that the Jurist al-Aṣīlī Used to Employ*
> Two dirhams of Byzantine gall nuts must be taken and ground finely; two dirhams of gum arabic must be taken and ground finely; each of them has to be placed separately. Then one dirham and a half of vitriol must be

11 Schopen, *Tinten und Tuschen* (cit. note 2), p. 125.

12 Translation by Lucia Raggetti; for the Arabic text, see MS Cairo Dār al-Kutub *ʿUlūm maʿāšiyya* 46, f. 29v. Schopen refers to this recipe erroneously by placing it at f. 30v and by considering it as the same preparation of the recipe of the ink used by al-Aṣīlī. Moreover, the title he gave — Recipe for an ink (*midād*) for paper and parchment — does not appear in the manuscript but it is more likely his own synthesis. Schopen, *Tinten und Tuschen* (cit. note 2), p. 104.

> taken, ground and placed separately. Then consider: if the ink is meant for paper, take the gum, put it in a mortar, pour twenty dirhams of water on it, grind the gum with water until it melts. Then add the ground gall nuts and the vitriol and grind everything with water until they are well mixed. Then put this in a vessel and write with it, if you want, immediately.
>
> If you want the ink to be for parchment, then the gum has to be one dirham and a half and the water two ounces, but you should not write with it until it has settled on the bottom.[13]

These two recipes for instant ink are very similar, especially in the part describing the ink for paper, as if the first preparation were a synthesis of the second. Interestingly, in the recipe of a good ink for paper the mention of vitriol is missing from the line where the ingredients are stated, possibly due to a mechanical error of omission, while its amount is given.

By comparing the part dealing with the ink for paper of these two recipes with that of Abū Ṭāhir, we can notice that they share the same procedures, the ingredient proportions (but not the actual figures) and even the textual structure, although the latter is more concise and some passages have been omitted. The obvious difference is that the vitriol is not mentioned among the ingredients and that the line stating the addition of vitriol to the solution is missing. It is my opinion, therefore, that the text of the recipe of Abū Ṭāhir is actually an incomplete recipe for an instant iron gall ink that, due to errors in the course of the textual transmission — again mechanical errors of omission, possibly due to the repetitive formulations typical of recipes — looks like a recipe for a plant ink. Two other elements support this hypothesis: among the few plant ink recipes that have been collected, all instruct the reader to extract the tannins by fermentation or cooking; none of the aforementioned recipes use gall nuts, mentioning instead other vegetal tannin sources, with a single exception that mixes damascene mulberries with a little extract of gall nuts.[14]

---

13 Translation by Lucia Raggetti; another translation is in Schopen, *Tinten und Tuschen* (cit. note 2), p. 103–104; for the Arabic text, see MS Cairo Dār al-Kutub *ʿUlūm maʿāšiyya* 46, f. 33r.

14 The recipe can be found in al-Marrākušī's *Kitāb al-azhār* and refers specifically to the "water of gall nuts"; Fani, *Le arti del libro* (cit. note 2), pp. 97–98; and al-Marrākušī, *Kitāb al-azhār* (cit. note 6), p. 85.

3 Case Study II — An Error and a Wrong Correction

In this case study, I analyse a recipe for iron gall ink that is part of the treatise *Tuḥaf al-ḫawāṣṣ* by al-Qalalūsī. The edition takes into account both the surviving manuscripts: MS Paris BnF Arabe 6844[15] and MS Rabat al-Ḫizāna al-Malikiyya 8998.[16] In the recipe to prepare a *midād* by cooking, we find a significant technical variant:

> *Another* Midād *Obtained by Cooking*
> One ounce (*ūqiyya*) and a quarter of gall nuts must be taken and ground finely in a mortar. One *raṭl* of fresh water must be poured on them and they must be left to rest for one night and one day. One ounce of **gum** [MS Paris BnF 6844] / **half ounce of vitriol** [MS Rabat al-Ḫizāna al-Malikiyya 8998] must be taken, ground in the mortar and the macerated gall nuts must be poured onto it, in a quantity that is enough to cover it, and it is left to rest for one day and one night until the water becomes red. At this point, the gall nuts are cooked on a low fire until a third has evaporated. They are sieved through a cloth in a clean vessel, then the gum is added while stirring; add to the mixture what you like until its tone satisfies you.[17]

The formula in the Paris manuscript is ineffective since the gum appears twice and the vitriol is missing. Most likely, one mention of the gum arabic was meant to be vitriol, but the scribe of this manuscript or of its model made a simple and trivial mistake while copying. The formula in the Rabat manuscript is, on the other hand, perfectly functional, with all the needed components mentioned, since the third sentence of the Rabat copy has "half ounce (*ūqiyya*) of vitriol" instead of "one ounce (*ūqiyya*) of gum" as found in the Paris manuscript. At first glance, a scholar might be tempted to correct the first formula with the second one, but by reading more carefully and reproducing them, it

15 The manuscript is composed of several units added to a core consisting of multiple texts copied by the same copyist, Muḥammad ibn Aḥmad al-Ġuraybī; the aforementioned treatise can be found in this core at ff. 112v–133v, it is dated at the end of *ḏū al-ḥiǧǧa* 986/ January 1578; Fani, *Le arti del libro* (cit. note 2), p. 134; Schopen, *Tinten und Tuschen* (cit. note 2), p. 27.

16 The severely damaged manuscript is dated Friday, middle of the month of *Ǧumādà al-ūlà* 993/ middle of May 1585. The name of the copyist is not given in the colophon; Fani, *Le arti del libro* (cit. note 2), p. 134; Schopen, *Tinten und Tuschen* (cit. note 2), p. 27.

17 See Fani, *Le arti del libro* (cit. note 2), pp. 138–139; and al-Qalalūsī, *Tuḥaf* (cit. note 7), pp. 22–23.

is clear that the first occurrence of gum arabic should be indeed gum arabic, while it is the second one that should have been vitriol. In fact, if gum arabic is added to macerating gall nuts it does not stop the process of extraction and the water will turn red-brown for the solubilization of the tannins (Fig. 7.1 B and B2). By contrast, adding vitriol to the gall nuts will result in a black solution and not a red one, as stated in the recipe, even if it is left for one day and one night (Fig. 7.1 A and A2).

Moreover, the last sentence of the recipe instructs the reader to add "what you like" in order to adjust its tone. But the intensity of the colour is not regulated by gum arabic, which influences the viscosity and the shine, but rather by vitriol. It thus makes more sense that the last ingredient added is vitriol and, since no amount is specified, the last sentence should be interpreted as "add to the formulation as much [vitriol] as you like until its tone satisfies you."

It is possible, then, that the scribe who wrote the Rabat manuscript (or its precursor) noticed the repeated ingredient in the passage he was copying and tried to correct the mistake: unfortunately, he corrected the wrong entry. In addition, in the version represented by the Rabat manuscript, the amount was reduced from one ounce (*ūqiyya*) to half of it, since the same quantity of vitriol and gall nuts for an ink prepared by maceration and decoction would have resulted in a formulation with an excess of iron and sulphuric acid that would probably have pierced the support. Both the corrections prove that the copyist, or the copyists, had a certain degree of competence but that they probably did not try out the recipe or else they would have noticed the inconsistency in the description.[18]

4 Case Study III — Sumac Ink: Black or Red?

In the third case study, I concentrate on three versions of a recipe for an ink made with vitriol, sumac and gum arabic that can be found in the *'Umdat al-kuttāb* by ibn Bādīs, in the *Muḫtara' fī funūn* by al-Malik al-Muẓaffar and in the *Kitāb al-azhār* by al-Marrākušī.

18 For the sake of clarity, I decided to simplify my reconstruction, although many possibilities could describe what really happened in the transmission of the recipe; unfortunately, the exact relation between the Paris and the Rabat manuscripts has not yet been investigated. Whatever the transmission might have been, it does not change the result: the corrections suggest a reasoned approach with some degree of competence in ink preparation.

FIGURE 7.1 Reproduction of the recipe by al-Qalalūsī: A) addition of vitriol to the solution of gall nuts (MS Rabat al-Ḫizāna al-Malikiyya 8998); B) addition of gum arabic in the solution of gall nuts (my reconstruction); A2) solution A after 24 hours; B2) solution B after 24 hours

The recipes in the *Muḫtaraʿ fī funūn* and in the *Kitāb al-azhār* are similar, although a different unit is used for the measuring of the gum arabic: three *raṭl* of fresh water are poured over a quarter *raṭl* of sumac from Sinǧār and the solution is left for two days in the sun. The residue is pounded and the solution strained. Then, for every *raṭl* of solution, five *ūqiyya* (*Muḫtaraʿ fī funūn*) or five *istār* (*Kitāb al-azhār*) of gum arabic are slowly added: one ounce (*ūqiyya*) / *istār* is added every day and then the solution is left in the sun. Finally, the vitriol is incorporated.[19]

The recipe in *ʿUmdat al-kuttāb* is different: one *raṭl* of sumac is cooked in six *raṭl* of water and, after the residue has been pounded and the solution filtered, five ounces of gum arabic are added to the solution. Once the gum dissolved, the vitriol is added.[20]

The word sumac (*summāq*) refers to a very useful plant: its fruits are used as spice and as a red dye, while its leaves and bark are traditionally employed in leather tanning. The recipes, however, do not specify which part of the plant should be used. All three recipes describe how the red colour will appear after the maceration or cooking of the sumac. This might be an indication of the use of fruits, however the red colour is also (though not always) mentioned when describing the liquid obtained from the maceration of gall nuts, thus it could refer to the colour of tannins, independently from the source. In both *ʿUmdat al-kuttāb* and *Muḫtaraʿ fī funūn* however, this ink description has been placed among those for coloured inks, while in *Kitāb al-azhār* it is recorded among those for black ones. The latter seems more logical to me, especially if leaves and bark were to be used, since they contain tannins that will react with the iron in the vitriol forming the black ferro-gallate complex.

I decided to test the recipe in *ʿUmdat al-kuttāb* using the fruits of sumac in order to see whether I could obtain a red ink. I put 30 g of sumac spice in 180 g of water and heated it until it reached boiling point. The recipe says to cook it "until the red appears" but the water became red right from the start of the cooking process and after 15 minutes I did not see any noticeable change in the colour of the water, although the spice gradually became pink-brown (Fig. 7.2 A). After straining the solution through a gauze, I added 12 g of gum arabic and waited for it to dissolve completely (this took about a day). I then added the vitriol and the liquid blackened immediately. After 6 g, the solution was very dark with some nuances of red-violet.

19 See in Fani, *Le arti del libro* (cit. note 2), pp. 63 and 90; Schopen, *Tinten und Tuschen* (cit. note 2), pp. 116–117; al-Marrākušī, *Kitāb al-azhār* (cit. note 6), p. 76.

20 See Schopen, *Tinten und Tuschen* (cit. note 2), p. 117; and Ibn Bādīs, *ʿUmdat al-kuttāb* (cit. note 4), p. 50.

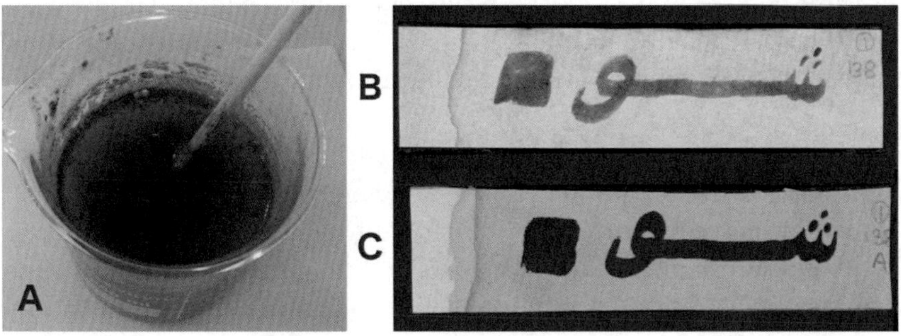

FIGURE 7.2 Sumac ink: A) extract from sumac fruit after boiling; B) sample of ink before ageing; C) sample of ink after ageing

This test proves that it is possible to obtain a black ink even from the fruit and therefore it is odd that both ibn Bādīs and al-Malik al-Muẓaffar inserted the recipe among coloured inks. It is possible that, knowing that a red dye can be obtained from the fruit of sumac, they overlooked the addition of vitriol and assumed it was a red ink. This demonstrates, however, that neither of them tested the recipe before assigning it to the right chapter, as they probably just took them over from their written sources. The position of this ink in the text speaks to the technical skills of the compilers, or rather their lack thereof.

Al-Marrākušī, on the other hand, was aware that after the addition of the vitriol the solution would become black; he specifically mentions this in his recipe. It is possible that he tried the ink before composing the text, thus demonstrating both his competence in the matter and a bibliographical attitude. It is still unclear, however, which part of the plant he is referring to. Considering that the colour of the solution made with the fruits was not stable and quickly faded into a red-brown, it is more likely that he intended the leaves and bark to be used (Fig. 7.2 B). In any case, even this ink turned extremely black after ageing (Fig. 7.2 C).

5 Case Study IV — Identification of Vitriols

In this case study, I initially focus on the recipe attributed to the physician Buḫtīšūʿ,[21] and then discuss the terms used for vitriols and the identification of

21 Dominique Sourdel, "Bukhtīshūʿ," in: *Encyclopaedia of Islam, New Edition* (cit. note 3), vol. I p. 1298. Schopen identifies him with Gibrīl b. Buḫtīšūʿ (d. 212/827), however, almost all the members of this family were physicians and bore similar names, thus the identification cannot be certain. See Schopen, *Tinten und Tuschen* (cit. note 2), p. 89.

the corresponding salts. The last sentence of the aforementioned recipe must have been particularly difficult to understand for later authors and compilers of other treatises, as it spawned three redactions with three different formulas.

> *Preparation of the Ink of Buḫtīšūʿ, the Physician*
> Take one *mikyāl* of ground gall nuts and eight *mikyāl* of water and cook them in a saucepan until one quarter of it has evaporated. Then, you remove it from the fire and leave it until it has cooled down. Strain it through a tight mesh rag and put it in a vessel. Add a sufficient amount of *zāǧ* (vitriol) and *qalqand* (green vitriol) and, in the same way, gum arabic, then write with it.[22]

> *Recipe for an Ink from the Buḫtīšūʿ the Physician, According to What Has Been Transmitted*
> Take one part of perfect gall nuts, pound them, pour on them eight parts of fresh water and cook them in a saucepan until a quarter of it has evaporated. Then, take it from the fire and leave it until it has cooled down. Then, strain it through a tight mesh rag and put it in a vessel. Then, add the required amount of *qalqand* (green vitriol), of good quality vitriol and the same amount of gum arabic. You write with it and it will be excellent, God willing.[23]

> *An Ink* (midād) *from Buḫtīšūʿ, Mentioned by al-Rāzī* [*MS Paris BnF Arabe 6844*] / *Another Ink* (midād) *Mentioned by al-Rāzī* [*MS Rabat al-Ḫizāna al-Malikiyya 8998*]
> Take one *mikyāl* of ground gall nuts and eight *mikyāl* of water. Cook them in a saucepan until a quarter of it has evaporated. Then, let it cool down and strain the formulation through a tight mesh rag. Add a sufficient amount of vitriol; once you have pounded it enough, do the same with the gum.[24]

---

22 See Fani, *Le arti del libro* (cit. note 2), p. 47; and Muḥammad ibn Zakariyyāʾ al-Rāzī, *Zīnat al-kataba*, ed. Luṭf Allāh al-Qārī, *ʿĀlam al-Maḫṭūṭāt wa-l-Nawādir*, 1432/2011:211–242, in particular pp. 225–226.

23 See Fani, *Le arti del libro* (cit. note 2), p. 93, Schopen, *Tinten und Tuschen* (cit. note 2), p. 89; and al-Marrākušī, *Kitāb al-azhār* (cit. note 6), p. 80.

24 See Fani, *Le arti del libro* (cit. note 2), p. 137; and al-Qalalūsī, *Tuḥaf* (cit. note 7), pp. 21–22.

Al-Rāzī wrote that gum arabic and two different types of vitriol, a generic one (*zāǧ*) and so-called *qalqand*, should be added at the end of the preparation, although no proportion or precise amount are given. Al-Marrākušī's version is slightly different, since in his recipe the gum arabic should be added in the same quantity as the vitriol. Then, as in the *Zīnat al kataba*, the same two types of vitriol are mentioned but the unit (*mikyāl*) was substituted with parts, possibly because that measure was no longer in use in Baghdad in the 13th century. Al-Qalalūsī, on the other hand, did not change the unit and, as al-Rāzī, understood that the gum had to be added to the preparation using the same method, but not necessarily in the same quantity as the vitriol. However, only one type of mineral, *zāǧ*, is mentioned. It is hard to guess the reason for which al-Qalalūsī did not mention *qalqant*, especially considering that he described it, in the last paragraph of the chapter on black inks, as one variety of vitriol. It remains possible, however, that the omission could be a change produced in the course of the transmission.

What is, then, the exact meaning of those terms? In the following paragraphs, I shall try to reconstruct a hypothesis of identification.

Zāǧ is the generic term for vitriol. It can be followed by adjectives specifying its colour or its place of origin.[25] The most frequently mentioned colours are yellow, green and white but it is extremely difficult to understand the chemical composition of these salts: they may refer to iron (III) sulphate (yellow), iron (II) sulphate (green) and zinc sulphate (white). Copper sulphate is blue in colour but there is no mention of blue vitriol in the treatises (Fig. 7.3).

Al-Rāzī explains some terms in his *Kitāb al-asrār* ('Book of Secrets'):[26] *qalqant/qalqand* corresponds to green vitriol (*zāǧ aḫḍar*), which, according to the procedure for its purification, seems to be copper (II) sulphate;[27] *qul-*

25 The regional indications are often too generic to precisely identify an area, as with Persian or Byzantine vitriol, and even when they are more precise, such as vitriol from Cyprus or from Kerman, several mines and consequently several types of vitriol have been found that correspond to the same area. Schopen, *Tinten und Tuschen* (cit. note 2), pp. 198–206. Moreover, the possibility that the names might actually be commercial labels must be considered. For example, is Cypriot vitriol a vitriol extracted from mines in Cyprus, or just a product typology similar to contemporary "Marseille soap," which identifies a product with recognisable characteristics but no real connection to the eponymous location?

26 Julius Ruska, *Al-Rāzī's buch Geheimnis der Geheimnisse* (Berlin: Julius Springer, 1937).

27 The procedure to prepare it: "Dissolve vitriol in water. Purify it. Throw on it copper filings and heat it until it is green. Purify it. Put it in a copper vessel. Dissolve it after you have put a half *dirham* of salt-ammoniac into ten *dirhams* of it."; Levey, *Medieval Arabic Bookmaking* (cit. note 2), p. 16, n. 88, Fani, *Le arti del libro* (cit. note 2), p. 226. According to this preparation method, it seems that the goal is a purification of an impure vitriol (probably a mixture of copper and iron sulphates) in an excess of copper solution to obtain a

FIGURE 7.3 Colours of various vitriols; top from left to right: pure iron (II) sulphate (green), copper (II) sulphate (blue) and zinc sulphate (white); bottom: mixture of iron (II) sulphate (yellow), iron (III) sulphate (green) and iron oxides (red)

quṭār/qalqaṭār corresponds to yellow vitriol (*zāǧ aṣfar*), although he does not explain how to obtain it; *qalqandīs/qalqadīs* refers to white vitriol (*zāǧ abyaḍ*) also called *šabb*, alum, which is potassium aluminium sulphate.[28]

Al-Marrākušī does not clarify any of the terms in his works, while al-Qalalūsī describes some of them in the last paragraph of the chapter on black inks. The text is unfortunately only partially legible due to damage to the manuscripts. He says that *qalqant* is green vitriol, while *qulquṭār*, also called shoemakers'

precipitate of metallic salts with a solubilization index lower of copper chloride, such as Fe (II) chloride. The solution will then contain copper ions, sulphuric acid and ammonium cations.

28 Procedure for its purification: "Take white pure alum. Dissolve and purify it. Distil vitriol and verdigris. Mix them with water of the purified alum and leave it in a beaker," Levey, *Medieval Arabic Bookmaking* (cit. note 2), pp. 16, n. 88, Fani, *Le arti del libro* (cit. note 2), p. 226.

FIGURE 7.4 Sample written with copper ink: on the left, application with a metallic pen; on the right, application with a reed

vitriol,[29] is the best, reacts quickly and does not leave mineral residues. He also informs us that Persian vitriol is clear and pure and, although a lacuna on the leaf prevents the reading of the colour, it does not have the hue of lapis lazuli (*lāzuwardī*). In another part of the text,[30] the *qulquṭār* is described as oxidized green vitriol, which, if we assume that green vitriol is iron (II) sulphate, should correspond to iron (III) sulphate, yellow in colour.

Green vitriol is the type most frequently mentioned in the recipes; thus, it is arguably the most effective. This would make sense if it referred to iron (II) sulphate, while al-Rāzī's identification with copper (II) sulphate is quite surprising.

In fact, copper (II) sulphate (blue in colour) does not form a black complex with gallic acid. To confirm the hypothesis, I conducted a simple experiment. I cooked 15 g of gall nuts in 30 g of water for ten minutes after boiling. I then progressively added pure copper (II) sulphate (0,3 g) to the still warm solution: it turned dark brown. I used it to write with a reed pen on handmade paper with no iron: as a result, the ink had a light brown colour (Fig. 7.4).

But what happens if some iron comes into contact with the solution? I used the same ink on the same paper but this time I wrote with a metallic pen, obtaining a very dark brown colour (Fig. 7.4, right side). This is due to the ionic exchange that happens when a solution with copper ions comes into contact with solid iron, caused by the different reactivity of the elements involved. In this case, the iron ions from the pen replace the copper ones in the ink

29 Pliny called copper sulphate (blue vitriol) "shoemaker black" (34.123); Pliny, *Natural History*, ed. and transl. by Harris Rackham, vol. 9: Books XXXIII–XXXV (The Loeb Classical Library, 394) (Cambridge, MA: Harvard Univ. Press, 1995).

30 See Fani, *Le arti del libro* (cit. note 2), p. 145; and al-Qalalūsī, *Tuḥaf* (cit. note 7), p. 28.

solution; the copper ions then solidify on the tip of the pen. Subsequently, the iron ions form the complex with the gallic acid in the solution, as is usual for iron gall inks. Since this exchange involves a small number of ions, the resulting colour is only dark brown and not black. It is important to remember, however, that the vitriols mined and then sold on the markets were never pure. In particular, iron sulphate was often obtained as a by-product of the mining of copper, thus traces or even consistent amounts of copper in iron sulphate and vice versa are not surprising. It is possible, then, that green vitriol referred to a more generic mixture of iron and copper sulphates, even taking into account the respective colours of the salts. The same consideration, however, cannot be made for the term *qalqand/qalqant*, at least not for the identification made by al-Rāzī in his in *Kitāb al-asrār*. It is possible nonetheless that the meaning of the term shifted over time and from place to place, towards a more general identification with green vitriol.

The presence of other metals in the vitriols has been seen as an opportunity in archaeometry, since the type of the impurities and their relative quantity can be used for provenance studies and to differentiate formulations of iron gall inks and mixed inks. This is particularly effective for the application of the fingerprint model[31] on the results obtained through X-Ray Fluorescence (XRF).

6 Case Study V — Non-vitriolic Iron Gall Ink

In the fifth case study, I analyse a type of iron gall ink made with iron filings or nails instead of the more common vitriol.

This ink was theorized as one of the first steps in the development of ferrogallic inks on the basis of evidence from manuscripts: in a recent study of Coptic manuscripts, several fragments dating from the 7th to the 10th century presented inks in which only iron and a small amount of manganese were identified using XRF, while copper and zinc, often associated with iron in vitriols,

31 The fingerprint model provides the ratio (W_i) between minor constituents, such as manganese, copper and zinc, and the concentration of iron, having excluded the contribution of the writing support; Oliver Hahn, "Analyses of Iron Gall and Carbon Inks by Means of X-ray Fluorescence Analysis: A Non-Destructive Approach in the Field of Archaeometry and Conservation Science," *Restaurator*, 2010, 31:41–64, pp. 47–48; Wolfgang Malzer, Oliver Hahn, Birgit Kanngießer, "A Fingerprint Model for Inhomogeneous Ink Paper Layer Systems Measured with Micro X-ray Fluorescence Analysis," *X-Ray Spectrometry*, 2004, 33:229–233.

were completely absent.[32] This can be due to an extremely pure vitriol but it is more likely that a different source of iron was employed. There are, however, no recipes from this time confirming the hypothesis.

Among the Arabic recipes, a small number mentions the use of iron filings or nails either on their own, or as an addition to vitriol. The former use is described by several authors: a recipe can be found in the *Kitāb al-azhār* by al-Marrākušī;[33] another one is in *Tuḥaf al-ḫawāṣṣ* by al-Qalalūsī[34] and also reformulated by an anonymous Sicilian scribe;[35] another is in Aḥmad ibn ʿIwaḍ ibn Muḥammad al-Maġribī's in *Kitāb qatf al-azhār fī ḫawāṣṣ al-maʿādin wa-l-aḥǧār* ('Book of the chosen flowers about the properties of metals and stones');[36] others are still used nowadays in sub-Saharan Africa, in particular in Kenya, Mali and Nigeria.[37] The distribution of the recipes shows that such ink typology was known and has remained in use until the present days, at least from the middle of the 13th century, the date of the oldest surviving attestation. In fact, the results of the analysis of the Coptic fragments suggest that the origin of this ink typology is older. It is also possible that the survival of this ink typology is due to the accessibility of the necessary ingredients, especially when vitriol, which was generally preferred for its ease of use, was not available on the market.

Although the polygenesis of this typology is very likely, the recipes are similar: the source of tannins, i.e. being gall nuts, bark from wild olive trees or mangroves, seed pods of the Egyptian thorn, or witchweed, is soaked in water, left to macerate or, more often, boiled with the iron pieces (in the form of filings, slags or rusty nails). In some recipes, an acid source, such as lemon juice, vinegar or date vinegar, is mentioned: it can be used to soak the iron before

32 The project and some of the results have been published in Tea Ghigo, Olivier Bonnerot, Oliver Hahn, Ira Rabin, Myriam Krutzsch, "An Attempt at a Systematic Study of Inks from Coptic manuscripts," *Manuscript Cultures*, 2018, 11:159–166, pp. 162–163.

33 See Fani, *Le arti del libro* (cit. note 2), p. 111, Schopen, *Tinten und Tuschen* (cit. note 2), p. 98 and al-Marrākušī, *Kitāb al-azhār* (cit. note 6), p. 102.

34 See Schopen, *Tinten und Tuschen* (cit. note 2), p. 124.

35 Eugenio Griffini, "Nuovi testi arabo-siculi" in *Centenario della nascita di Michele Amari*, (Palermo: Virzì, 1910), Vol. 1 pp. 443–448.

36 Manfred Ullmann, *Die Natur- und Geheimwissenschaften im Islam* (Leiden: Brill 1972), p. 137; for this recipe see also Schopen, *Tinten und Tuschen* (cit. note 2), p. 99.

37 Ahmed Parkar collected a recipe from Kenya made with mangrove wood, lemon juice and rusty nails, which was reproduced during the course of this study, see Colini, *From Recipes to Material Analysis* (cit. note 1) pp. 94–95, Eva Brozowsky replicated inks from Mali, including some using local plants and nails, while examples from Nigeria using nails and local plants and date vinegar can be found in Biddle, see Biddle, *Inks of Northern Nigeria* (cit. note 2), pp. 14–16.

mixing with the tannins or added after the boiling of the plants and the filings. The first procedure results in the production of iron ions and their reduction from iron (III) ions to iron (II) ions (Fe^{3+} to Fe^{2+}), which react with gallic acid forming the iron (II) gallate complex. It is interesting to note that, although the black pigment is due to the iron (III) gallate complex, the latter is poorly water soluble, while the transparent iron (II) gallate complex is highly soluble, meaning that the black complex tends to precipitate if a huge amount of it is formed in the inkwell. It is therefore easier and more effective to write with an ink consisting of little precipitate, i.e. with part of the molecules in an iron (II) gallate state, which turns completely black only after being applied to the writing support; that is, when the iron (II) gallate complex is oxidized by the oxygen in the air into iron (III) gallate. For this reason, the efforts of the ink makers focused on procedures that led to obtaining iron (II) ions, or else the reaction would be in favour of iron (III) gallate complex, with the ink then resulting in a dispersion rather than a solution.[38] The addition of the acid after boiling the tannins may serve a similar purpose, since an acidic environment prevents the oxidation reactions turning iron (II) ions into iron (III) ions (Fe^{2+} into Fe^{3+}). But tannic extracts have a pH ranging from two to three, comparable to that of vinegar and lemon juice, therefore the change in acidity of the solution obtained by the addition of the acid should not be relevant. However, acid other than tannins act to produce iron (II) ions (Fe^{2+}) without binding iron, resulting in a better stoichiometric relation between iron and tannins. In fact, the two reproduced recipes made with the addition of acid (from contemporary Kenya and al-Qalalūsī's treatise, Fig. 7.5 A and 7.5 B respectively) have a darker colour after ageing than the one made without (by al-Marrākušī, Fig. 7.5 C).

Interestingly, al-Marrākušī was aware of the resulting colour, as the title of the preparation, "Production of an ink that comes out in the *mahrī* colour"[39] shows; moreover, it is included in the chapter on coloured inks, not in the chapter on black *ḥibr* (iron gall inks). This is further proof of his competence and possibly of the effort he put into testing his recipes before recording them.

The results obtained by XRF analysis of the samples made with the reproduced inks are comparable to those observed on manuscripts, thus

38 It is possible that the vitriol was preferred over the iron filings to make iron gall inks because of the difficulty and length in producing a good ink with iron oxides, although the preparation is more stable in time since no sulphuric acid is present.

39 According to Schopen, it is the brown-black colour typical of camels, Schopen, *Tinten und Tuschen* (cit. note 2), p. 98; Fani, however, translates it as the colour of the fruit of the colocynth, Fani, *Le arti del libro* (cit. note 2), p. 111; for the Arabic, see al-Marrākušī, *Kitāb al-azhār* (cit. note 6), p. 102.

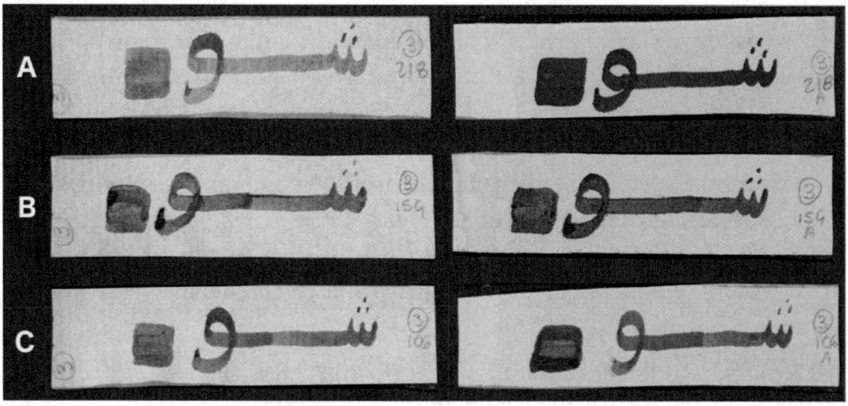

FIGURE 7.5 Samples written with iron gall inks made with iron filings before (on the left) and after (on the right) ageing: A) recipe from contemporary Kenya, ink made with mangrove bark, lemon juice and rusty iron filings; B) recipe by al-Qalalūsī, ink made with bark, vinegar and iron filings; C) recipe by al-Marrākušī, ink made with gall nuts, iron filings and gum arabic

confirming that the analysed Coptic inks may have been produced with iron filings, slags or nails.

7 Case Study VI — Mixed Inks, a Problematic Typology

For the last case study, I am considering an entire typology of inks: the mixed inks. They have been described by Zerdoun as such: "Ce sont par exemple des encres au carbone auxquelles le préparateur a ajouté des extraits aqueux de produits tannants ou des sels métalliques; ce sont également des encres métallo-galliques dans lesquelles on a incorporé du noir de fumée."[40] Gacek says of them, "mixed inks consist of ingredients used traditionally in both carbon and iron gall inks (carbon based and tanning agents)."[41] Although it is certainly difficult to describe all the hues that characterize this class of inks, the aforementioned definitions are quite generic. At least two main subgroups should be distinguished: carbon ink combined with plant ink, and carbon ink mixed with iron gall ink.[42] The former was known and used: for example, the

40 Monique Zerdoun Bat-Yehouda, *Les encres noires au Moyen Âge (jusq'à 1600)* (Paris: CNRS Éditions, 2003), p. 20.

41 Adam Gacek, *Arabic Manuscripts. A Vademecum for Readers* (Leiden: Brill 2009), p. 133.

42 Only Schopen mentioned this division; Schopen, *Tinten und Tuschen* (cit. note 2), p. 14. A third group might also be recognized, carbon ink with the addition of metallic salts, but since the colour is given only by the carbon component, we consider it to be a subgroup

Jewish philosopher Maimonides (d. 1204 CE) described it as the preferred ink for the writing of phylacteries.[43] The mixed inks, combining carbon and iron gall inks, were also very popular in the Arab world since they combined the intense black colour and stability of the carbon ink with the indelibility of ferrogallic inks.

At times, the recipes instruct the reader to mix a complete ink from one typology with additional ingredients from a different typology: for example, Persian ink (carbon-based) with tannins, thus obtaining a mixed carbon-plant ink. But more often a list of ingredients of different ink typologies would be given, resulting in a "natural born" mixed ink. Moreover, the proportions of the ingredients could vary greatly, resulting in inks with very different behaviours and characteristics. In general, the authors of the treatises and the compilers do not give them a specific name but call them *ḥibr* (iron gall inks) or *midād* (carbon inks) depending on how close their characteristics are to the first or the second category. The same is true for the chapter in which they are included. The only exception is al-Marrākušī, who grouped the mixed inks of both subgroups in a chapter dedicated to simple liquid *midād* and *midād murakkab* (composite ink).

The first attestations of recipes for mixed carbon-plant ink date from between the late 9th century and the early 10th century, since both ar-Rāzī and ibn Muqla[44] described this typology in their treatises. This typology is not very representative, its formulas being only six per cent of the total, but parallels can be found with the Jewish tradition, especially with the so-called Maimonides' ink, which is a variant of the one described by ibn Muqla.[45] The mixed carbon-iron gall inks represent 14 per cent of the formulas collected. The majority of the recipes can be found in al-Marrākušī and are ascribed to

of the class of carbon inks. Recipes with such a description can be found in Greco-Roman antiquity (by Dioscorides, 1st c.) and Jewish tradition from the 2nd c.; Zerdoun, *Les encres noires* (cit. note 30), pp. 79–81 and 105–110. No such recipe has been found among the Arabic texts.

43 Zerdoun, *Les encres noires* (cit. note 30), pp. 111–116.

44 Dominique Sourdel, "Ibn Muḳla," in: *Encyclopaedia of Islam, New Edition* (cit. note 3), vol. III pp. 886b–887a; translations of his recipe in Schopen, *Tinten und Tuschen* (cit. note 2), p. 130 and Ahmed Mousa, *Zur Geschichte der Islamischen Buchmalerei in Aegypten*, Būlāq Cairo: Government Press 1931, p. 27.

45 Zerdoun, *Les encres noires* (cit. note 30), p. 111.

him or his contemporaries, probably evidence of the increasing popularity of this medium at that time.[46]

Together, they represent 20 per cent of the formulas but their use is rarely attested in manuscripts.[47] The results obtained by analysing some of the reproduced recipes of mixed ink show that the protocol currently in use is seldom successful in identifying them.

In general, Near infrared reflectography (NIR) is a quick and perfect method for dealing with a pure class ink, since carbon, plant, and iron gall inks have very distinct optical properties. However, no unequivocal identification of mixed inks seems possible, since a high amount of carbon masks the presence of other components when illuminated with NIR light.[48] Ultra-violet reflectography (UV) can offer complementary data thanks to the tannins' property of quenching fluorescence, which enhances the contrast between a fluorescing background and the text, but here, too, the presence of carbon on the surface of the ink will partially or even completely mask the result, depending on how damaged the ink is.

Raman spectroscopy presents the cleanest and the most straightforward method for identifying carbon and iron gall inks and is therefore well suited to documenting a mixture of both. Unfortunately, despite the recent development of portable Raman spectrometers, the analysis of black inks still often requires a bench instrument or the extraction of samples in addition to trained personnel. Moreover, the detection limits of the different components (iron gall inks and carbon) need to be further tested. Furthermore, Raman measurements of plant inks and of mixed carbon-plant inks yield no conclusive spectra of tannins with lasers in the visible wavelength range (VIS). The situation is somewhat better when the excitation wavelength is shifted to the

---

46 In the recipe for the *ḥibr* used by Ḫālid ibn Barmak (d. 165/782) al-Marrākušī mentions a variant introduced by the vizier Yaḥyā b. Ḫālid (d. 190/806), who was said to add myrobalan and oil soot to his father's preparation. This, then, would be the first attestation of such a mixed ink, see Fani, *Le arti del libro* (cit. note 2), pp. 100–101. The author of the *Kitāb al-azhār*, however, is cautious with the attribution and, in fact, he reported the news as something that "was told in Baghdad in the year 649 H / 1251 CE," hinting that he had no written or more reliable source to confirm the rumours and perhaps that an authority was needed to empower and justify the use of the new style of ink.

47 For a summary of the results achieved and the techniques used, see: Claudia Colini, Oliver Hahn, Olivier Bonnerot, Simon Steger, Zina Cohen, Tea Ghigo, Thomas Christiansen, Marina Bicchieri, Paola Biocca, Myriam Krutzsch, Ira Rabin, "The Quest for the Mixed Inks," *Manuscript cultures*, 2018, 11:43–50.

48 Ralf Mrusek, Robert Fuchs, Doris Oltrogge, "Spektrale Fenster zur Vergangenheit. Ein neues Reflektographieverfahren zur Untersuchung von Buchmalerei und historischem Schriftgut," *Naturwissenschaften*, 1995, 82:68–79, p. 72.

NIR region, but even in this case strong fluorescence of organic molecules considerably disturbs the spectrum. Surface-Enhanced-Raman-Spectroscopy (SERS) may overcome this difficulty; indeed, it has already been applied successfully to the analysis of modern paints and dyes.[49] This is, however, a micro-invasive technique that, depending on the selected substrate, requires sampling. Attempts have been made, for cultural heritage purposes, to reduce the sample amount to a minimum and to optimize it. However, an optimized substrate and procedure for SERS on tannins and mixed inks must still be defined.[50] Similar problems concern the application of Infra-Red Spectroscopy (FTIR) for the detection of tannins. The use of methods such as Laser Ablation for micro-sampling, however, could encourage the application of FTIR Spectroscopy in transmission. The accuracy of the results obtained using Mass Spectroscopy (MS), in particular Atmospheric Solids Analysis Probe Mass Spectoscopy (ASAP-MS), suggests that it could be an even more promising technique, but it still requires micro-sampling.[51] The samples made with the aforementioned recipes are currently been used to test these methods.

8 Conclusions

The sentence "I tried it and it is really good," in all its declinations, can be found at the beginning or end of many recipes. This could easily be a *topos* of this technical literature and does not necessarily mean that the formulation was tested;[52] it is only meant to give prestige and authority to the recipe. But we know that a few authors and compilers, such as al-Marrākušī, enacted the recipes or evaluated their feasibility by analogy.[53] These elements, in addition to the easily observable fluidity of the tradition, suggest a "hands on" approach to the transmission of such texts, where practical experimentation was, if not common, at least considered an important and validating element. It is apparent both when the writer was copying a treatise from a specific *Vorlage* (case

49 Federica Pozzi, Marco Leona "Surface-enhancement Raman Spectroscopy in Art and Archaeology," *Journal of Raman Spectroscopy*, 2016, 47:67–77.

50 Colini *et al., The quest* (cit. note 47), p. 49.

51 Tea Ghigo, Ira Rabin, Paola Buzi, "Black Egyptian inks in Late Antiquity: new insights on their manufacture and use", *Archaeological and Anthropological Sciences* 2020, 12:70 (1–10), pp. 5 and 11–12.

52 For example, no information is given regarding the ineffectiveness of recipes, although some of the formulations, when tested, did not give the expected results.

53 Al-Marrākušī uses this word to explain the process of testing a recipe on the basis of similar recipes that proved successful.

study II), or when he was in the process of importing recipes from a treatise to form a new one (case studies III and V).

Nowadays, the replication of ink recipes may fulfil several functions.

It proves helpful for finding mistakes and problematic points in the text of the recipes, such as erroneous corrections of terms or missing information (case study I–II), but also for understanding the order and organization, or lack thereof, of the treatise (case study III). It can also clarify the meaning of terms and expressions (case studies III and IV) and, in a broader view, it helps in assessing the real technical expertise of the writers, or at least the effort and attention that they put into writing their texts (case studies II, III, IV and V). Through the replication it is possible to prepare samples that, once analysed, can support the observations made on manuscripts by comparing their analytical results (case study V). Furthermore, the effectiveness of the current analytical protocols and equipment can be evaluated on the basis of which ingredients and ink typologies can be identified through their application (case study VI).

The case studies show the key role of the replication of recipes for both the Humanities and the Natural Sciences. The interdisciplinarity required to effectively carry out and comprehend the procedures clearly highlights the need of increasing the cooperation between scholars of different academic branches.

CHAPTER 8

Ordinary Inks and Incredible Tricks in al-ʿIrāqī's *ʿUyūn al-ḥaqāʾiq*

Lucia Raggetti

Abstract

The *Kitāb ʿuyūn al-ḥaqāʾiq wa-īḍāḥ al-ṭarāʾiq* ('The best of true facts and the explanation of their ways') was composed in the 13th century by Abū al-Qāsim al-ʿIrāqī, best known for his alchemical works. This peculiar handbook counts 30 chapters and includes many different streams of tradition: pseudo-Platonic magic, medicine, pharmacology, sleight of hand, and crafts. This chapter focuses on the recipes for coloured metallic inks and invisible ones (chapters 18 and 23) and provides an edition and a commented English translation of these sections. The kind of edition proposed here — a 'laboratory-edition' — is devised as a specific tool for interdisciplinary research on premodern science and technology and as preparatory work for the replication of these recipes.

Keywords

metallic inks – invisible inks – fluid traditions – technical literature – al-ʿIrāqī – replication – 'laboratory-edition'

1 The Author, the Text and Its Transmission

The 13th-century alchemist Abū al-Qāsim al-ʿIrāqī (7th/13th century)* — also known as al-Sīmāwī ('the practitioner of natural magic') — composed a peculiar collection of 30 chapters, entitled *Kitāb ʿuyūn al-ḥaqāʾiq wa-īḍāḥ al-ṭarāʾiq*

* This publication is part of the research project *Alchemy in the Making: From Ancient Babylonia via Graeco-Roman Egypt into the Byzantine, Syriac, and Arabic Traditions*, acronym *AlchemEast*. The *AlchemEast* project has received funding from the European Research Council (ERC) under the European Union's Horizon 2020 research and innovation programme (G.A. 724914).

('The best of true facts and the explanation of their ways'), dealing with all sorts of tricks, deceptions, wonders, and the specialists in these fields.[1] Different scholars have pointed out the role of this text in the history of magic and its importance as an indirect witness to the pseudo-Platonic *Kitāb al-nawāmīs* ('Book of natural laws,' the *Liber Anegueminis* in the Western tradition).[2] The magical components also include the preparation of talismans and the invocation of spiritual entities. Magic, however, is only one of the streams of tradition that converge in the *ʿUyūn al-ḥaqāʾiq*.

This work also includes a technical component often expressed in the explanation of many illusionistic tricks and in the instructions for different preparations. Some materials are connected to specific groups of specialized tricksters and have a parallel attestation in al-Ǧawbarī's *Kitāb al-muḫtār fī kašf al-asrār* ('Anthology on the unveiling of secrets'), a 13th-century handbook that unveils the tricks of street frauds.[3] In the *Kašf al-asrār*, the dupes are arranged on the

1 See Eric J. Holmyard, "Abu' l-Qāsim al-ʿIrāqī," *Isis*, 1926, 3:403–426. For the alchemical works of al-ʿIrāqī, see also *Kitāb al-ʿilm al-muktasab fī zirāʿat adh-dhahab* (*Book of Knowledge Acquired Concerning the Cultivation of Gold*), ed. and trans. Eric J. Holmyard (Paris: Librarie Orientaliste Paul Geuthner, 1923); and the *Book of the Seven Climes* (*Kitāb al-aqālīm al-ṣabʿah*), focusing on alchemical illustrations. A digital copy has been made available by the British Library (MS London BL Add. 23390, ff. 50v–87v), <http://www.qdl.qa/en/archive/81055/vdc_100023587816.0x000002> (last accessed 9 April 2020); this manuscript is also described and discussed in a post in the British Museum blog by Bink Hallum and Marcel Marée, see <https://blog.britishmuseum.org/a-medieval-alchemical-book-reveals-new-secrets/> (last accessed 9 April 2020).

2 See Liana Saif, "The Cows and the Bees: Arabic Sources and Parallels for Pseudo-Plato's *Liber Vaccae* (*Kitāb al-nawāmīs*)," *Journal of the Warburg and Courtauld Institute*, 2016, 79:1–48; for the Mediaeval Wester tradition of the text, see Maaike van der Lugt, "'Abominable Mixtures': The *Liber Vaccae* in the Medieval West, or the Dangers and Attractions of Natural Magic," *Traditio*, 2009, 64:229–277; Paolo Scopelliti and Abdelsattar Chaouech, *Liber Anegueminis. "Il libro della vacca" dello pseudo-Abū Zayd Ḥunayn ibn Isḥāq ibn Sulaymān ibn Ayyūb al-ʿIbādī* (Milano: Mimesis, 2006); and Manuela Höglmeier, *Al-Ǧawbarī und sein* Kašf al-asrār — *ein Sittenbild des Gauners im arabisch-islamichen Mittelalter* (*7./13. Jahrhundert*) (Berlin: Klaus Schwarz Verlag, 2006), p. 396.

3 For the Arabic text and a thorough commentary, see Höglmeier, *Al-Ǧawbarī und sein* Kašf al-asrār (cit. note 1); for the French translation, see ʿAbd al-Rahmâne al-Djawbarî, *Le Voile arraché. L'autre visage de l'Islam*, 2 vols, translated by René R. Khawam (Paris: Phébus, 1979). On the one hand, the two authors might have tapped into the same sources to produce independent works that partially overlap. On the other, it is possible that al-ʿIrāqī used the *Kašf al-asrār* as source, which, at that time, must have been a very recent addition to technical literature in Arabic. Although al-Ǧawbarī arranged the materials differently, he treated subjects that also found a place in the *ʿUyūn al-ḥaqāʾiq*: soporifics, tricks of the conjurers (with writing) and of the astrologers, stratagems to discover thieves. For the parallel attestations in the *ʿUyūn al-ḥaqāʾiq*, see Appendix II and Höglmeier, *Al-Ǧawbarī und sein* Kašf al-asrār (cit. note 1), pp. 346, 233, 214 and 245–250. Tricks contemplating

basis of the different professional groups (alchemists, pharmacists, food merchants) who perpetrate them. In the *ʿUyūn al-ḥaqāʾiq*, however, this approach is limited to a few chapters and many other examples of technical expertise are detached from a specific professional context. They are presented rather as amusing technical tricks of dexterity (from magic boxes to bent swords to simulate stabbing). Another stream of technical traditions is represented by the medical components, dealing with simple drugs, occult properties of natural objects and the constitution of man. The result of this complex merging of sources is a handbook that exists in the intersection between natural magic, technical knowledge, and sleight of hand.

The author added a brief introduction to the text, in which he declares the reasons that brought him to the composition of the book and a general recapitulation of its contents. The last remark of the introduction seems to refer to an encoding of the text carried out by the author himself, although the terminology usually refers to writing and calligraphic styles.[4]

> Abū al-Qāsim ibn Aḥmad ibn Muḥammad known as al-ʿIrāqī said: "When I saw that the stratagems of the greater part of the natural things had been made manifest among many groups of tricksters, but they could not achieve anything from the true facts without any claim or science, I decided to write this book and to entitle it 'The best of true facts and the explanation of their ways'.

different writing practices are included in the section devoted to the conjurers, the section on the secrets of writing (*asrār al-kitāba*) exclusively deals with ways to erase writing from different supports, see Höglmeier, *Al-Ǧawbarī und sein* Kašf al-asrār (cit. note 1), pp. 303–307. The 13th century also saw the composition of al-Iskandarī's (fl. 640 H/1243 CE) *Al-ḥiyal al-bābiliyya*. Chapter 14 of this text treats several procedures to encode writing with different cryptographic techniques, invisible inks arranged by the substance that makes them appear, the erasure of writing from papyrus and parchment, and how to dye leaves in different colours. See al-Ḥasan ibn Muḥammad al-Iskandarī, *Al-ḥiyal al-bābiliyya li-l-ḫizāna al-kāmiliyya* (Al-Iskandariyya: Maktabat al-Iskandariyya, Markaz Dirāsāt al-Ḥiḍārat al-Islāmiyya, 1439/Alexandria: Islamic Civilization Studies Center, 2018). Later, al-Zarḫūrī wrote a handbook to instruct the tricksters, see Lucia Raggetti, "*Cum Grano Salis*. Arabic Ink Recipes in their Historical and Literary context," *Journal of Islamic Manuscripts*, 2016, 7/3:294–338, pp. 328–329. This text is also divided into 30 chapters and its author, though the chronology is not certain, was contemporary to al-ʿIrāqī, possibly one generation older. For an anthology of translated passages from these works, see also Lucia Raggetti, *Un coniglio nel turbante. Intrattenimento e inganno nella scienza arabo-islamica* (Milano: Editrice Bibliografica, 2021).

4 The textual tradition of the introduction is very stable, with only minor variants that do not affect the meaning, which allowed me to give a single translation. For the variety and use of secret alphabets, see, for instance, the *Kitāb mabāhiǧ al-aʿlām fī manāhiǧ al-aqlām* ('Book of the delights of the signs in the methods of the pens') by al-Bisṭāmī (d. 858 H/1454 CE) as attested in MS Leiden Or. 14.121. See pp. 48–49 of Jan Just Witkam's *Inventory of the Oriental Manuscripts of the Library of the University of Leiden*, <http://www.islamicmanuscripts.info/inventories/leiden/or15000.pdf> (last accessed 1 March 2020).

It deals with some of the stratagems (*ḥiyal*) from the *nawāmīs* (lit., '[natural] laws'), incendiary preparations (*maḥārīq*), fumigations (*al-daḫan*), fermentations (*al-taʿāfīn*), soporifics (*al-marāqid*), astrological incantations (*al-nāringǧāt*), concealments (*al-aḫfāʾ*), illusionistic tricks (*al-dakk*), stratagems (*al-ḥīla*), the occult properties of stones, minerals, plant and animals (*ḥawāṣṣ al-maʿdan wa-l-nabāt wa-l-ḥayawān*), and the natural composition of man (*tarkīb al-insān*) and what is specific for it at every moment.

So, I divided it into 30 chapters, each dealing with a witty artifice for the one who wishes to understand its explanation and meaning, and among these there are also the secrets that should not be unveiled. We noted it down in *rayḥānī* [writing] style ⟨and adorned it in *ʿIrāqī* [writing] style, MS Princeton Garrett 544H⟩ so that only the competent one can access to them".

The introduction is followed by a list of the 30 chapters and their respective titles with a summary of their contents.[5]

Eight different witnesses to the text were collected for this study — seven manuscripts and a lithographic edition — and represent the basis for the critical work on the text. An introduction by the author is attested in all the witnesses and is regularly followed by a list of the 30 chapter headings. Some of them make use of a secret alphabet to encode key technical information (for instance, the name of an ingredient or its precise quantity) and in two of them one can even find a *legenda* to interpret these signs. Curiously, in the two copies that sport a *legenda*, the secret alphabet is not specifically used to encode significant bits of the text.

1.1 (*P*) *MS Princeton Garrett 544H (150 ff.)*[6]

A date written at the end of the text by the copyist who produced the whole manuscript indicates that the copy was completed on the 7th Ḏū al-Ḥiǧǧa 1274 H / 19th July 1858 CE. The manuscript is written in a cursive *nasḫ*, chapter headings and the incipits of their subdivisions are rubricated. A secret alphabet is used to encode the technical details of different procedures. If we consider, however, the instances in which the corresponding letters of the Arabic alphabet are given in *inter lineam* — by what seems to be the same hand as that of the copyist, using the same ink of the main text — the association

5 See Appendix 1.

6 A digital copy of the manuscript is available at <http://pudl.princeton.edu/objects/qz20ss55t#page/297/mode/1up> (last accessed 9 April 2019).

between the letters and the signs of the secret alphabet is not consistent. This manuscript also features a few drawings of magical signs, diagrams, and tables that summarize the text.

1.2 (*B*) MS *Berlin Wetzstein II 1375 (70 ff.)*[7]

The manuscript is written in a cursive *nasḫ*, the chapter headings and the incipits of their subdivisions are rubricated. Along with a few drawings of magical signs, this copy includes vivid illustrations of some spontaneously generated creatures described in the fourth chapter and the schematic but detailed drawings of some tools to perform tricks (magical boxes, bent swords, etc.) that illustrate an additional section on the sleight-of-hand (*šaʿbaḏa*), wedged between Ch. 8 and Ch. 9. Some specific technical information is not encoded with a secret alphabet, however in the relevant passages the letters are written in their isolated form. Here, the list of chapters is given a layout usually reserved for poetry, with a clear division between the two halves of the line. Some folia are annotated in the margins by the hand of a reader who added parallel recipes and procedures, either collated from a different copy or collected from other materials at his disposal.

1.3 (*D*) MS *Dublin Chester Beatty Ar. 4019 (68 ff.)*[8]

This undated manuscript is written by two different hands, a main one responsible for the greater part of the text, along with a second one that intervenes in a few instances between Ch. 23 and Ch. 25. The chapter and paragraph headings are rubricated in the parts written by the main hand, while they remain black for the second one, usually in bold and sometimes marked by a super linear stroke. Like the Berlin manuscript and in the same position, this witness includes an additional section on different tricks of legerdemain (*šaʿwaḏa*) accompanied by illustrations of the different devices involved in the tricks, though depicted in a different order. Crucial portions of the procedures are, in some cases, encoded in a secret alphabet, for which the manuscript does not provide a *legenda*.

1.4 (*T*) MS *Toronto Fischer Library 142 (122 ff.)*[9]

The manuscript is written in a cursive *nasḫ*, the chapter headings and the incipits of their subdivisions are rubricated. The manuscript features a few

7 Wilhelm Ahlwardt, *Verzeichniss der arabischen Handschriften der Königlichen Bibliothek zu Berlin*, vol. 5 (Berlin: Asher, 1893), p. 99 No. 5567.

8 A digital copy of the manuscript is available at <https://viewer.cbl.ie/viewer/object/Ar_4019/1/> (last accessed 15 April 2020).

9 A digital copy of the manuscript is available at <https://archive.org/details/uyunalhaqaiqwaidoounse> (last accessed 9 April 2020).

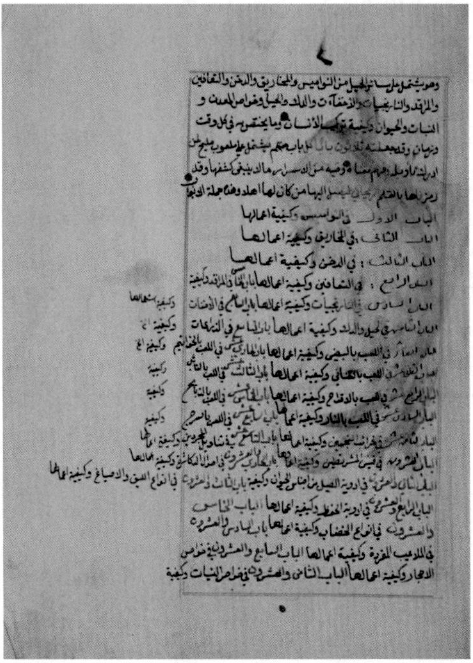

FIGURE 8.1
List of chapters following the incipit of the *ʿUyūn al-ḥaqāʾiq*, MS Toronto Fischer Library 142, p. 2

drawings of magical signs, diagrams, and tables that summarize the text. It is paginated with Arabic numbers and the same hand added another table of contents with page numbers on one of the blank leaves at the end of the manuscript. The same hand also added a *legenda* for the secret alphabet used in the manuscript — again, on a blank leaf after the end of the text; here, the rubrications are made with a different ink, purple rather than red — although no part of the text is actually encoded. The colophon tells that the copy was completed in the month of Ramaḍān 1285 H / December 1868 CE (Fig. 8.1).

1.5 (*K*) *MS Jeddah King Saud Library 6230 (72 ff.)*[10]

The manuscript is written in a very cursive *nasḫ*, the chapter headings and the incipits of their subdivisions are rubricated. The manuscript features a few drawings of magical signs, diagrams, and tables that summarize the text. The copyist occasionally annotated the margins with corrections and additions to the text, though some marginal annotations could also be ascribed to a different hand. Before the colophon, there is a *legenda* for the secret alphabet used in a number of cases to encode specific technical information. The colophon tells that the copy was completed in the year 1272 H / 1855–56 CE.

10 A digital copy of the manuscript is available at <https://al-mostafa.info/data/arabic/depot/gap.php?file=m017532.pdf> (last accessed 9 April 2020).

1.6 (*L*) MS *London British Library Add. 23390 (ff. 50v–87v)*[11]

This is a multiple-text manuscript matching the *Mechanics* by Hero of Alexandria (*Kitāb fī rafʿ al-ašiyāʾ al-ṯaqīla*, 'On the lifting of heavy things')[12] with the *ʿUyūn al-ḥaqāʾiq*, which produces an interesting combination of different technical texts. The text of the *Mechanics* is enriched with numerous diagrams representing the various machines; these are associated with rubricated progressive numbers expressed by the numerical value of Arabic letters. Other rubrications added to the diagrams indicate their different components. The text of the *ʿUyūn al-ḥaqāʾiq* does not contain any diagrams, but several blank spaces suggest that they were part of the initial plan. The original colophon has been erased and replaced with a 19th-century version (f. 87v). An ownership note on f. 1r, however, marks a *terminus ante quem* at the year 1020 H/ 1611 CE. The manuscript was copied by an expert *nasḫ* hand, the rubrications in the diagrams might have been added by a different one.

1.7 (*La*) MS *London British Library Or. 3751 (ff. 1v–28r)*

The first part of this multiple-text manuscript contains an abridgement of al-ʿIrāqī's *ʿUyūn al-ḥaqāʾiq* (*Fawāʾid min kitāb ʿUyūn al-ḥaqāʾiq*), also the other two texts in the collection are abridgements of medical and alchemical works. The title page is missing, a blank leaf at the beginning has the *legenda* of a secret alphabet and the title of the work written upside down, probably from a different hand. The text is written in a regular *nasḫ*, the chapter and paragraph headings are rubricated, the margins are ample and often filled with annotations and corrections, probably from the same hand.

1.8 (*C*) *Cairo lithographic edition (48 pp.)*[13]

The title page of the lithographic copy of the *ʿUyūn al-ḥaqāʾiq* sports a frame divided into two rectangular areas. In the upper one, there is a circular medallion containing a long version of the title and the name of the author with the eulogies of the case. The lower one contains four lines informing us that this edition was printed at the expense of Mister ʿAlī ʿAbd al-Ḥamīd al-Kutubī and

11 A digital copy of the manuscript is available at <https://www.qdl.qa/en/archive/81055/vdc_100022551545.0x000001> (last accessed 23 April 2020).

12 For this text, see Carra de Vaux, *Les Mécaniques ou l'Élévateur de Héron d'Alexandrie, publiées pour la première fois sur la version arabe de Qostâ ibn Lûqà et traduites en français* (Paris: Leroux, 1894).

13 A digital copy of the lithograph is available at <https://gallica.bnf.fr/ark:/12148/bpt6k9106144f/f5.item.zoom> (last accessed 9 April 2020).

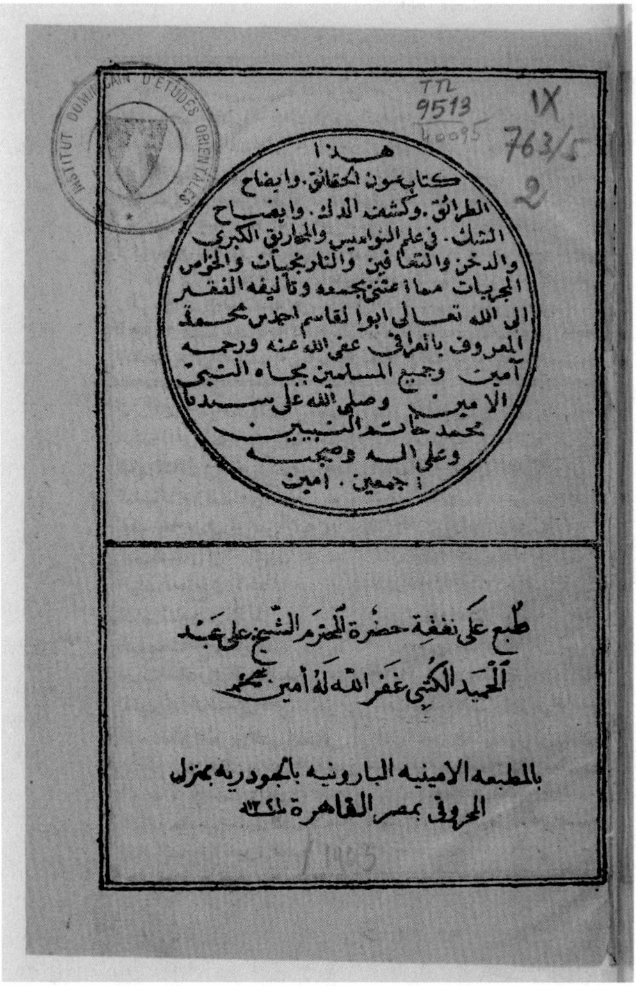

FIGURE 8.2 Title page from the lithographic edition of the *'Uyūn al-ḥaqā'iq* (Cairo, 1321 H/1903 CE)

printed at the *Maṭba'a Bārūniyya* in Cairo,[14] in the year 1321 H/1903 CE. On the following page, the frame is divided into an upper rectangular area including the introduction, and two columns underneath with the list of chapters. On all remaining pages, the text is framed in a rectangle defined by a double line. The incipit of the different chapters is marked by flowered brackets (Fig. 8.2).

14 For this printing press, see Martin H. Custers, *Ibadi publishing activities in the east and in the west c. 1880–1960s, An attempt to an inventory, with references to related recent publications* (Maastricht: Custers, 2006).

2 Ordinary Inks and Incredible Tricks

Writing plays a role in different practices described in al-ʿIrāqī's work and writing, along with inks, stands out as an example of the technical vein in the *ʿUyūn al-ḥaqāʾiq*. Not all ink types are represented in the text, which devotes an entire chapter (Ch. 23) to coloured metallic inks and paints, and part of another one to invisible inks (Ch. 18). Compared to technical handbooks on ink making, the selection of the materials included in this treatise is limited and peculiar.[15]

For the edition of the text — with a practical approach to the fluidity of the tradition — the more inclusive version has been chosen for the main text — i.e. MS Princeton Garrett 544H — and Appendix II gives an overview of the attestation of the recipes in the different witnesses of the manuscript tradition.[16] Although it is not possible to define stemmatic relations among the manuscript witnesses, it is still possible to detect some proximity between some of them. The manuscripts T and K, for instance, share a mechanical mistake: recipe nos. 8–10 and 28 are matched with the wrong rubric; these recipes carry in fact the title of the following entry.

The edition presented here is a small philological experiment that I will call a 'laboratory-edition'; that is to say, it is an edition devised for interdisciplinary use and for collaboration between philologist and chemist. This edition is oriented by its prospective readership and is designed to make the text and its variants accessible, especially to those without direct access to primary sources. Thus, the variant readings are translated and, whenever necessary, commented upon.[17] Regarding the variants to be included in the apparatus: the 'laboratory-edition' operates a selection and only those focused on the technical aspects of the text are included in the apparatus, leaving aside small orthographical and linguistic variants that do not imply a technical difference. This kind of edition is the first step towards the replication of recipes and provides the material information for setting the research questions that replication may find an

15 Following the order of technical treatises on ink making, metallic inks will be dealt with before the invisible ones, although the *ʿUyūn al-ḥaqāʾiq* treats them, respectively, in Ch. 23 and Ch. 18. The title of Ch. 23 mentions metallic inks and dyes (*al-liyaq wa-l-aṣbāġ*) but also includes instructions for preparing paints (*dihān* or *adhān*) from the same mixtures meant for inks; whereas Ch. 25 is entirely devoted to dyestuffs (*al-ḫiḍābāt*) for hair and beard.

16 The summary of the contents has been prepared taking into account the complete manuscript tradition.

17 Every edition is the result of a compromise between three parties: the text, the editor, and the imagined readership. See Francisco Rico, "Los Quizotes de Hartzenbusch," in *Juan Eugenio Hartzenbusch, 1806/2006*, edited by M. Amores (Madrid: Centro para la edición de los clásicos españoles), pp. 199–220, in particular pp. 203 and 209.

answer to.[18] The apparatus also includes a section on parallel attestations of the recipe (*loci similes*) in different treatises on ink making.[19]

2.1 *Coloured Metallic Inks*

This broad approach to the collection of sources leads the author to include an entire chapter on metallic inks and coloured paints obtained from the same mineral or metallic compound.[20] For other research, there had already been occasion to establish the fluidity of the text and its contents in the relative stable frame of the 30 chapters that compose the book, and the chapter on inks confirms the impression.[21] The attestation and distribution of the recipes in the different witnesses shows significant variations (Appendix II). The chapter structure, however, remains constant: opened with a recipe for preparing

18 See Lucia Raggetti, "Inks as Instruments of Writing: Ibn al-Ǧazarī's *Book on the Art of Penmanship*," *Journal of Islamic Manuscripts*, 2019, 10/2:201–239.

19 In the section of the apparatus reserved for parallel attestations of a recipe (*loci similes*), the references to handbooks on ink making and other relevant texts are referred to in an abbreviated form: 'al-Marrākušī' for Muḥammad ibn Maymūn ibn 'Imrān al-Marrākušī, "Kitāb al-azhār fī 'amal al-aḥbār li-Muḥammad ibn Maymūn ibn 'Imrān al-Marrākušī," *Zeitschrift für Geschichte der Arabisch-Islamischen Wissenschaften*, 2001, 14:103–106; 'Ibn Bādīs' for al-Mu'izz ibn Bādīs al-Tamīmī al-Ṣanhāǧī, *'Umdat al-kuttāb wa-uddat ḏawī al-albāb. Fīhi ṣifat al-ḫaṭṭ wa-l-aqlām wa-l-midād wa-l-liyaq wa-l-ḥibr wa-l-asbāġ wa-ālat al-taǧlīd*, edited by Naǧīb Mā'il al-Harawī and 'Iṣām Makkīya (Mašhad: Maǧma' al-Buḥūṯ al-Islāmīya, 1409 / 1988 H); 'al-Qalalūsī' for Abū Bakr Muḥammad ibn Muḥammad al-Qalalūsī al-Andalusī, *Tuḥaf Al-Jawāṣṣ Fī Turaf Al-Jawāṣṣ* (*Las galanduras de la nobleza en lo tocante a los conocimientos más delicados*), edited by Hossam Ahmed Mokhtar El-Abbady (Alexandria: Maktabat al-Iskandarīya, 2007); 'al-Rāzī' for Muḥammad ibn Zakariyyā' al-Rāzī, *Zīnat al-kataba*, ed. Luṭf Allāh al-Qārī, *'Ālam al-Maḫṭūṭāt wa-l-Nawādir*, 1432/2011, 16/2:211–242; 'Fani' for Sara Fani, Le arti del libro secondo le fonti arabe originali. I ricettari arabi per la fabbricazione degli inchiostri (sec. ix–xiii): loro importanza per una corretta valutazione e conservazione del patrimonio manoscritto (PhD Diss., Università degli Studi di Napoli "L'Orientale," 2013); '*Cum Grano Salis*' for Raggetti, *Cum Grano Salis* (cit. note 2), in particular for al-Zarḫūrī's *Zahr al-basātīn*; and 'Siggel, *Decknamen*' for Alfred Siggel, *Arabisch-Deutsches Wörterbuch der Stoffe* (Berlin: Akademie Verlag, 1950); and 'al-Iskandarī' for al-Iskandarī, *Al-ḥiyal al-bābiliyya* (cit. note 3).

20 The themes of colours and writing are also present in other sections of the book. Invisible inks are treated in Chapter 18 among the tricks of those who dupe people with written messages that suddenly appear or disappear, while Chapter 25 deals with dyeing substances. See Appendix I. The name *līqa* refers to inks by extension, being this a wad of unspun silk, wool or cotton placed in the inkwell's neck to prevent the ink from being spilled when the pen is dipped in it, see Adam Gacek, *Arabic Manuscripts. A Vademecum for Readers* (Leiden/Boston: Brill, 2009).

21 In particular, the reading of the fourth chapter on wondrous fermentations in the different witnesses reveals a high degree of variance in the wording and contents of the various witnesses to the *'Uyūn al-ḥaqā'iq*.

the gum arabic that is needed for the preparations of all the inks; a number of coloured metallic inks followed by a technical consideration on the composition of colours; and a final part with recipes dealing with chrysography. Manuscript P is the more inclusive version chosen for the main text: the recipes for metallic inks are understandably more numerous and this is the only text to include a procedure for cutting and applying gold leaves (no. 34). This cluster of recipes for coloured metallic inks (nos. 12–27) was probably inserted before the recipe preceding the remark on the composition of colours ('wood ink,' here recipe no. 28). The idea that this insertion and its position are deliberate is supported by the fact that the recipe for the 'wood ink' is first partially copied before this additional cluster of recipes, and then copied in its entirety after it. The concise style of this particular cluster of recipes, suggests that it represents an addition in this specific copy rather than an omission from the others. The order of the recipes, however, does not apparently follow the one adopted by any other treatise on ink making in particular. Once the position of this additional cluster of recipe has been defined in the frame of the textual tradition of the *'Uyūn al-ḥaqāʾiq*, this material remains relevant from the technical point of view and therefore finds its way in the main text of the 'laboratory-edition.'

Chapter 23: on [metallic] inks (*liyaq*) and the manner of their operations

الباب الثالث والعشرون في انوع الليق وكيفية اعمالها

الليق] الليق والاصباغ '[metallic] inks and dyes,' *add.* K C

[1]

The Wise said: for the one who wishes to prepare [metallic] inks and paints, it is necessary to begin first with the preparation of chips of white gum arabic.

قال الحكيم ينبغي لمن اراد عمل الليق والاصباغ ان يتبدئ اولا بتدبير الصمغ العربي الابيض المعقرب

Take the preferred quantity of it, crush it finely, soak it in pure water in a glass vessel.

يأخذ منه ما اختار فيدقه ناعما ويبله بالماء الصافي في اناء زجاج

Then, add three parts of water for each part it, close the opening of the vessel and hang it in the sun for a whole day.

ثم يعطي لكل جزء منه ٣ اجزاء من الماء ويسد رأس الاناء ويعلقه في الشمس يوما كاملا

Then, shake it until [the components] blend one with the other, and agitate it until it has settled.

Then, take the required quantity needed to fix metallic inks, dyes, and paints.

When it dries and melted sandarac is applied over it, then this paint will not be removed, even if washed with water.

ثم يحرك حتى يختلط بعضه ببعض ويترك حتى يركد

ثم يأخذ منه بقدر ما يحتاج لاصلاح الليق والاصباغ والادهان

فان جف ودهن من فوقه السندروس المحلول فانه لا يعود يزول ذلك الدهن ولو غسله بالماء

الحكيم] المفيد (Wise → Instructive), K C

عمل] علم (to prepare, lit. ‘making’ → science) P D

الابيض] الابيض المحلول (white → white and dissolved), *add.* B T K C

المعقرب] وهو ان يأخذ من الصمغ العربي الابيض المعقرب *add.* (in chips → that is white gum arabic in chips), B T K C

ويبله بالماء الصافي] ويحل ثم ينقع في الماء (soak it in pure water → dissolved and then soaked in water), B; وتخله ثم تبله بالماء (→ sifted and then boiled in water), T K; تديف الى الماء (→ diluted with water), L; وتنقعه بالماء (→ immersed in water), C

بالماء الصافي] بعد ان تغليه على النار في اناء نظيف الى ان يصلب في الاناء (in pure water → after having been boiled on the fire in a clean vessel until it has hardened in the vessel), *add.* C

اناء] وعاء (vessel → receptacle, either a different kind of recipient or a different name for the same one) B D T K L La

ثم يعطي] ثم ينقع به الصمغ في اناء زجاج او من مزجج (Then, add → then let the gum arabic macerate with it in a glass vessel or in a glazed one) C

يركد] يركض P; يمتزج (it has settled → has mixed) C

السندروس] الصندروس P

غسله] علقه (washed → suspended) La

فان...بالماء] واذا جف اقلب عليه الماء (When it dries and melted sandarac is applied over it, then this pain will not be removed, even if washed with water → Once it has dried, pour some water on it) D

Loci similes: *Cum Grano Salis*, p. 335.

[2] Cinnabar ink

ليقة زنجفر

Take some red cinnabar, grind it finely, then rinse it with the water of sour pomegranate seeds, pour water over it and rinse well, and purify it after leaving it for one hour, until [the suspension] has settled.

يؤخذ من الزنجفر الاحمر ويسحق ناعما ثم يصول بماء حب الرمان الحامض وتقلب عليه ماء وتغسله غسلا جيدا وتصفيه بعد ان تتركه ساعة حتى يركد

Then, grind it either in an impermeable or on a polished permeable stone slab, add water gradually and grind it until it cannot absorb any more water and is similar to *ḥarīra* soup.

ثم يسحق على صلاية مانع او صلاية ناعمة ملسا واسقيه بالماء قليلا قليلا وانت تسحقه الى ان لا يعود يشرب ماء ويبقي كالحريرة

Then, add the dissolved gum arabic and pound it vigorously until it is absorbed into the substance.

فحينئذ تلقي عليه الصمغ المحلول واسحقه به قويا حتى انه يدخل في جسمه

If you wish to make it into an ink, pour this on a washed silk wad inside a glass jar and write with it. If, instead, you want to use this for paints, use a hair brush to spread it onto the images you wish and then the leaf will be coloured in a precious red, and you should know this.

فان اردته ليقة تزلته على ليقة حرير مغسولة في حق زجاج واكتب ما اردت وان اردته للدهان فمشيه بقلم الشعر على ما اخترت من الصور فانه يصبغ الورق احمرا ثمانيا فاعلمه

ليقة زنجفر] صفة ليقة زنجفر T K; صفة ليقة زنجفرية L C
الاحمر] (red) *om.* B
حب] (grain) *om.* B C; خشب (→ wood) La
يركد] يرسب (has settled → sinks to the bottom) C
غسلا...ثم يسحق] *om.* D
صلاية مانع] (impermeable [...] stone slab) *om.* T K L C
صلاية ناعمة] رخام املس ناعما (levigated and permeable stone slab → levigated and permeable marble) B; صلاية ملسا (→ levigated stone slab) C
واسقيه بالماء قليلا قليلا] (add water gradually) *om.* C; بالماء قليلا (→ with a bit of water) L C
وانت تسحقه الى ان لا يعود يشرب ماء] (grind it until it cannot absorb any more water) *om.* C

كالحريرة [كالخرزة (similar to *ḥarīra* soup → like a bid) B T; الحديد الناعمة (→ fine iron) L; كانه الاسفداج (→ as if it were white lead) C; مثل الحرير (→ like silk) D Ba. *Ḥarīra* is a soup of flour cooked with grease, gravy, or milk; *ḥarīra* may also mean silk cloth.

تلقي عليه الصمغ المحلول واسحقه به قويا حتى] (Then, add the dissolved gum arabic and pound it vigorously until) *om.* K T; لخر يدخل خمسة في بعض (→ So that five times as much is absorbed?) C

قويا] (vigorously) *om.* B

فان اردته ليقة] ان اردته كالدهن ليقة (If you wish to make it into an ink → If you want it to be like a paint [in the form of] ink) L C

ليقة حرير] ليقة احمر (silk wad → red wad) B

في حق] في جرة (jar → vase) B

زجاج] نظيف (glass→ clean glass) *add.* C

انه يصبغ الورق احمرا ثمانيا] (then the leaf will be coloured in a precious red → then the leaf will be coloured red) *om.* B T K L C

فانه...فاعلمه] *om.* D

Loci similes: several similar recipes for a cinnabar red ink can be found in the treatises on ink making, though not in such a detailed form. In these recipes, usually pomegranate water is substituted with gall nut water. See Fani, pp. 64, 66, and 127 (cinnabar ground in a stone pounder) and 104–106.

[3] Arsenic ink

Take some golden yellow arsenic (orpiment), pulverize it, sift it, and grind it with water on an impermeable stone slab until it does not absorb any more water; pour the dissolved gum arabic onto it until you are satisfied with it and store it away for the moment when you may need it, either for writing or for the paints, it will be excellent.

ليقة زرنيخ

يؤخد زرنيخ اصفر ذهبي يطحن وينخل ثم يسحق على صلاية مانع بالماء حتى لا يعود يشرب شيئا والق عليه الصمغ المحلول الى حين يرضيك وارفعه لحاجتك اما للكتابة واما للدهان بلاغه

واما للدهان بلاغه] واما للدهان فاعلم ذلك (or for the paints, it will be excellent → or for the paints, and you should know this) T K L La; او للدهان (→ or for the paints) C

Loci similes: see Fani, p. 108 (yellow ink); al-Marrākušī, p. 98.

[4] [Another ink]
If yellow arsenic (orpiment) is not available, grind Iraqi white lead, pour freshly plucked saffron and gum water over it, then do with it whatever you wish.

[ليقة اخرى]
اذا اعدم الزرنيخ الاصفر اسحق الاسفيداج العراقي والق عليه الزعفران الجنوي والق عليه ماء الصمغ وافعل به ما شئت

ليقة اخرى صفراء [ليقة اخرى (another ink → another yellow ink) B; *om.* P

الجنوي] *al-ǧannawī*. This adjective can be interpreted as the *nisba* for the city of Genoa. The 'saffron of Genoa' is not attested as a label for a certain product that is associated with this place of origin (such as the vitriol of Cyprus or the *terra sigillata* from Lemnos). The only other occurrence I could find is in a 14th-century treatise on the art of writing, mainly calligraphy. The editor specifies in a footnote that this variety of saffron comes from Italy (*Īṭāliyā*) and it is renowned for its abundant juice and the beauty of its colour. He also suggests checking Ibn al-Bayṭār in the Cairo lithographic edition. A perusal of the relevant entry on saffron does not reveal the use of this adjective. The only passage with relevant information is at the beginning, when Ibn al-Bayṭār reports Dioscorides' description of the different geographical varieties of the plant with their specific names and properties. One of the varieties mentioned is said to be typical of a country called *Aṭūliyā*, while another variety from Sicily is said to have a lot of juice and a beautiful colour and is used as a dye by the people of *Anṭāliyā*. This lexicographical direction is not particularly productive or reliable, so the other option may be an adjective from the participial form (*ǧannā*) of the verb *ǧanā* ('to pluck fruits and flowers from a plant'). See Ibn al-Bayṭār, *Kitāb al-ǧāmiʿ li-mufradāt al-adwiya wa-l-aġdiya* (Cairo: Maktabat al-Mutanabbī), p. 126; and Ḥusayn ibn Yāsīn ibn Muḥammad al-Kātib, *Laḥmat al-muḫtaṭif fī ṣināʿat al-ḫaṭṭ al-ṣalaf* (Kuwayt: Mu'assasat al-Kuwayt li-l-taqaddum al-ʿilmī, 1992), p. 73, note 232.

Loci similes: *Cum Grano Salis*, p. 331 (no. 124)

[5] Green ink
Take some well-ground yellow arsenic (orpiment), add a quarter dirham of indigo for each *miṯqāl* of it [yellow arsenic]; grind it until you like the green hue of its colour, pour the dissolved gum onto it and do with it whatever you wish, either for writing or for [the preparation of] paints.

ليقة خضراء
يؤخذ الزرنيخ الاصفر المسحوق ناعما ويلقي على كل مثقال منه ربع درهم نيل هندي واسحقه الى حين يعجبك لونه في الخضرة ونزل عليه الصمغ المحلول وافعل به ما اردت للكتابة وللدهان

صفة ليقة خضراء [ليقة خضراء TKC

دهان وغير ذلك [وللدهان (for [the preparation of] paints → for paints and other than this) *add.* C
Here K adds a recipe for preparing 'gold water' (صفة ماء الذهب):

تأخذ كبريت وشبا ايضا اجزاء اسواء اسحقهما حتى يختلطوا ثم اجعلهما في قدرة واغليها على النار غلوتين واتركه حتى يبرد واعمله مثل الفلوس وجففه في الظل ثم اسحق منه قدر الحاجة بخل عتيق وصمغ عربي واكتب به ما شئت فانه يكون على لون الذهب

'Take some sulphur and an equal quantity of white alum, grind them until they mix; then put it in a pan and boil it twice on the fire, leave it until it has cooled, give it the shape of small coins and let it dry in the shade. Then, grind the amount you need with old vinegar and gum arabic and write whatever you wish with it and it will have the colour of gold.'

Loci similes: see Fani, p. 130; al-Marrākušī, p. 128.

[6] Verdigris ink
Grind Iraqi verdigris with wine vinegar well, then pour the gum onto it and do with it what you wish.

ليقة زنجاري
يسحق الزنجار العراقي بخل الخمر سحقا ناعما ثم يلقي عليه الصمغ وافعل به ما اردت

الصمغ] ماء الصمغ (gum arabic → water of gum arabic) B D T K L La C
ما اردت] ما شئت من كتابة او دهان فانه يأتي كما تحب وتختار (what you wish → what you wish, either for writing or for paints, and it will be as you like and prefer) *add.* D T K L; ما تريد فانه يأتي كما تحب (what you want and it will go as you like) *add.* C

[7] Pistachio ink
This is when you take the verdigris ink and pour on it two *dāniq* of freshly plucked saffron, and this will improve its colour and satisfy you.

ليقة فستقية
وهو ان تأخذ الليقة الزنجاري ويلقي عليها قدر دانقين زعفران جنوي فانه يحسن لونه ويرضيك

دانقين] دانق D (two *dāniq* → one *dāniq*)

Loci similes: although several recipes for a pistachio ink are preserved in the technical literature, none is based on the same ingredients, see Fani, pp. 65, 67, 109 and 131; and al-Marrākušī, pp. 100 and 129.

[8] White ink
Take some Iraqi white lead, grind it finely with the water of white gum arabic until it [the product] pleases you and its whiteness has improved, then do with it whatever you wish, either an ink or a paint.

ليقة بيضاء
تأخذ الاسفيداج العراقي وتسحقه ناعما بماء الصمغ الابيض الى حين يعجبك ويجود بياضه فاصنع به ما شيت اما ليقة واما دهان

فاصنع به ما شئت اما ليقة واما دهان] (then do with it whatever you wish, either an ink or a paint) *om.* B

[9] Lapis lazuli ink

This is when a quarter dirham of indigo must be added for every dirham of white lead, then grind it well in a thin impermeable mortar until something similar to lapis lazuli remains; then use it as an ink, or for the basic preparation (*biṭāna*) for paints with lapis lazuli.

ليقة لازوردي
وهو ان يلقي على كل درهم اسفيداج ربع درهم نيله هندي واسحقه جيدا في صلاية ناعم مانع حين يبقي شبيه اللازورد واعمل منه ليقة او بطانة للدهان باللازورد

ناعم] (thin) *om.* D

اللازورد] وعمل عليه الصمغ المحلول (lapis lazuli → and add to it the dissolved gum arabic) *add.* L; واعمل منه ليقة او بطانة للدهان باللازورد (then use it as an ink, or for the core preparation of paints with lapis lazuli) *om.* B

[10] Wine ink

Grind the red lac, add to it a piece of indigo and prepare an ink or a paint.

ليقة خمري
يسحق اللك الحمراء واطرح عليها قطعة نيلة هندية واجعله ليقة او دهن

اللك الحمراء] الحمرة (red lac → red dye) B; اللكة الحمراء (→ red lac) T; اكليل الملك الحمرة (→ red sweet clover) K; اللك الاحمر (→ red lac, with a different spelling) L

Loci similes: Ibn Bādīs includes a recipe with the same label in his treatise, but the ingredients are very different (gall nuts water and arsenic), see Ibn Bādīs, p. 49.

[11] Turtle-dove ink

Add a drop of ink (*ḥibr*) for every dirham of sericon or a little bit of dissolved indigo.

ليقة فاختي
الق قطرة من الحبر على كل درهم من السيلقون او شيئا يسيرا من النيلة الهندية المحلولة

المحلولة] المبلولة (dissolved → moistened) D

السيلقون] السيرقون B, the transliteration of the Greek name for the alchemical red tincture (σηιxόν, 'sērikón') could be written in Arabic in both ways; it may indicate 'cinnabar' or refer to other red substances (see recipe No. 13), Siggel, *Decknamen*, p. 82.

[12] Rosy ink

Take some cinnabar and white lead, grind them in a vessel and add the gum; if you wish to obtain an intense pink, add more of the red [component], whereas if you want a lighter [pink], add more of the white [component].

ليقة وردي
يؤخذ من الزنجفر والاسفيداج واسحقهما في اناء وصمغهما فان اردت الوردي عميق فاجعل الاحمر اكثر وان اردته صافي اعمل الابيض اكثر

Loci similes: al-Marrākušī, p. 92; Fani, pp. 106–107; Ibn Bādīs, p. 59.

[13] Orange ink

Take some sericon and grind it. Dissolved gum must be added to it, then one can write with it and it will be nice; if you want a lighter hue, take some yellow arsenic (orpiment) and add it to the red cinnabar.

ليقة نارنجي
يؤخذ السيلقون يسحق ويعمل عليه الصمغ المحلول ويكتب به يجيء مليح وان اردته صافي خذ من الزرنيخ الاصفر ويعمل عليه الزنجفر الاحمر

السيلقون] السيرقون B (see recipe no. 11).

Loci similes: Ibn Bādīs, p. 64.

[14] Honey ink

Take one part of [carbon] ink (*midād*), one part of cinnabar, add the gum and write what you wish.

ليقة عسلي
يؤخذ جزء من المداد وجزء من الزنجفر ويصمغ ويكتب ما اراد

[15] Silver ink

Take common silver leaves; do with them the same as you would do with gold — as it will come later — and add gum to them.

صفة ليقة الفضية
يؤخذ اوراق الفضة المتاع الطلي افعل بهم كما تفعل بالذهب كما سياتي واعمل عليهم الصمغ

[16] Black vitriolic ink
Take some gall nuts, crush them, macerate them in water, write with this on a leaf and leave it until it fades away. Then, take some good vitriol from Cyprus, grind it well, mix it with water and wash the leaf in it, and the writing will appear black.

الليقة السوداء الزاجية
يؤخذ العفص يسحق وينقع بالماء ويكتب به في الورق وتتركه الى ان ينشق ثم تأخذ الزاج القبرصي المليح تستحقه ناعم وتذوبه في الماء وتغمس الورقة فيه فتظهر الكتابة سوداء

Loci similes: see recipe no. 39 in ch. 18 of the *ʿUyūn al-ḥaqāʾiq*; Fani, p. 152; al-Qalalūsī, p. 36.

[17] Golden ink
Take some copper filings and golden marcasite, grind the filings and wash the marcasite white; then grind them on a stone slab, and mix it with gum arabic and then one can write with it.

ليقة ذهبية
يؤخذ برادة النحاس ومرقشيتا ذهبية تسحق البرادة وتغسل المرقشيتا ابيضا ويسحقوا على صلاية ويضاف لهم صمغ عربي ويكتب به

[18] Ashen ink
White lead, cinnabar, a bit of Egyptian ink (*midād miṣrī*), and it will be excellent.

ليقة رمادية
اسفيداج وزنجفر وقليل مداد مصري فانه غاية

[19] Violet ink
Take — one or more pieces [?] — white lead, indigo, and cinnabar. They must be ground finely and the gum has to be mixed with them.

ليقة بنفسجية
يؤخذ الاسفيداج والنيلة والزنجفر جزء وجزؤن [؟] يسحقوا ناعما ويضاف اليهم الصمغ

Loci similes: although recipes for violet ink are attested in the technical literature, none enumerates these same ingredients, see Fani, pp. 132 and 146; al-Marrākušī, p. 130.

[20] Turquoise ink
Iraqi verdigris, white lead and a bit of saffron, prepare as described before.

ليقة فيروزجية
زنجار عراقي واسفيداج وقليل زعفران ويعمل كما تقدم

[21] Peony ink
Verdigris and saffron.

ليقة شقائقي
زنجار وزعفران

[22] Clay ink
It is made from the mentioned red ochre together with [egg?] white, and the colours will result from increasing [the proportion of one ingredient].

ليقة سجي
يقوم من المغرة المذكورة مع البياض وعلى
قدر الزيادة تكون الالوان

Loci similes: *Cum Grano Salis*, p. 330 (no. 118)

[23] Lentil ink
From saffron and white [egg? lead?].

ليقة عدسي
من زعفران والبياض

[24] Abbasid ink
From black [dye], red ochre in flakes.

ليقة عباسي
من السواد المغرة السطيحة

[25] Ivory ink
From arsenic with [egg? lead?] white.

ليقة عاجي
من الزرنيخ مع البياض

[26] Golden ink
From arsenic and lac.

ليقة ذهبي
من الزرنيخ واللك

[27] Preparation of the golden, silver, copper, and lead ink and of every metal with the splendour of the two [gold and silver?], then it will take its colour
Its preparation: take a fine [powder of] touchstone, mix it with gum arabic, and write with it; once it has dried, rub and burnish with gold and the writing will appear golden — or, if you do it with silver, [the writing] will become silver, or, apart from these two, any metal you wish, and you should know this.

صفة الليقة الفضية والذهبية والنحاسية
والرصاصية وكل معدن بجليتهما فتصير
على لونه

وصفتها ان تأخذ المحك ناعم وتخلطه
بالصمغ العربي وتكتب به فاذا جف
ونشف صقلته بالذهب تطلع الكتابة
ذهبية او بالفضة تصير فضية او منهما
شئت من المعادن فافهم ذلك

Loci similes: see *Cum Grano Salis*, p. 330 (no. 117); *Art of Penmanship*, recipe no. 21.

[28] Wood ink

Grind some red arsenic (realgar) finely, pour the closest thing to the most suitable/closest/lowest quality ink (*ḥibr*) that there is onto it; or take the red [arsenic], add the yellow [one] and the [carbon] ink (*midād*), and all this after the addition of gum.

You should know that the colours are produced from each other, when you add one to the other, depending on the difference of the colours, so be aware of this.

ليقة عودية

اسحق الزرنيخ الاحمر سحقا ناعما والقي عليه ادني ما يكون من الحبر او خذ الاحمر والق عليه الاصفر والمداد وذلك جميعة بعد التصميغ

واعلم ان جميع الالوان تتولد بعضها من بعض اذا القيت على بعضها بعض باختلاف الالوان فاعلمه

او خذ الاحمر والق عليه الاصفر والمداد وذلك جميعة بعد التصميغ] او اجمع بين الاحمر والاصفر والمعادن ويسحق جيد ويضاف اليه ماء الصمغ المقدم ذكره واستعمله (or take the red [vitriol], add the yellow [one] and the [carbon] ink, and all this after the addition of gum → or mix the red, the yellow and the minerals, grind it, mix the gum water mentioned before and use it) B; *om.* L

التصميغ] الصمغ (addition of gum → gum) D

اذا القيت على بعضها بعض باختلاف الالوان] (when you add one to the other, depending on the difference of the colours) *om.* B C;

باختلاف الالوان] باختلاف الاوزان (depending on the difference of the colours → depending on the difference of weights) K; باختلاف الاوزان وذلك كله بعد التصميغ (→ depending on of the difference of weight and all of this after the addition of gum) L

[29] Section on the melting of gold

When you wish this, take a blue Chinese bowl, smooth and fine, throw ten leaves of the finest quality Egyptian gold into it; pour half a dirham of bee honey onto this, gradually make everything into a single leaf, until it melts; then wash the sweetness of the honey from it with water, leave it to sink to the bottom, filter it from the water, pour a *ḥarrūba* [dry measure] of gum arabic moistened with water onto it and write with it.

فصل في حل الذهب

اذا اردت ذلك فخذ زبدية صيني زرقاء ملسا ناعمة واطرح في وسطها عشرة اوراق ذهب مصري عال والق عليه وزن نصف درهم عسل نحل وتمرس الجميع في ورقة قليلة قليلة حتى ينحل واغسل عنه حلاوة العسل بالماء واتركه يرسب وصفي عنه الماء والق عليه وزن خروبة صمغ عربي مبلول بماء واكتب به

When it dries, leave it for one hour and then burnish it with hematite or with onyx, from which a burnisher is made, and its colour will appear and will be as you like and as it pleases you.

فاذا جف فاتركه ساعة واصقله بحجر الصرف او بجزع يكون قدعمل منه مصقلة فانه يظهر لونه ويأتي كما تحب وترضي

فصل في حل الذهب] (Section on the melting of golf) *om.* D
زرقاء] (blue) *om.* L C
مصري] (Egyptian) *om.* D
نصف درهم] نصف درهم ورقة ورقة (half dirham → half dirham leaf by leaf) *add.* D
وزن نصف درهم] وزن خروبة (half dirham → the weight one *ḫarrūba*) T K C; ووزن ورق (→ the weight of leaves/a leaf?) L
حلاوة] جزء (sweetness → part) L; *om.* C
واتركه يرسب وصفي عنه الماء والق عليه وزن خروبة] (leave it to sink to the bottom, filter it from the water, pour a *ḫarrūba* onto it) *om.* L
مبلول] محلول (moistened → dissolved) D

Loci similes: see Fani, p. 71; *Art of Penmanship*, recipe no. 19.

[30] Melting of all the metals

If you want this, take a whetstone, rub it on any stone available or any metal you wish, then this will gradually dissolve [by filing] until nothing of this substance is left.
Then add gum arabic in drops, burnish it when it has dried and the colour will appear.

حل جميع المعادن
اذا اردت ذلك فخذ حجر المحك وحك عليه اي حجر كان واي معدن شئت فانه ينحل اولا باول حتى لا يبقي من ذلك الجسم شيء
ثم قطر عليه الصمغ واكتب به فاذا جف اصقله فانه يظهر لون

اي حجر كان] (any stone available) *om.* D T K L
الجسم] الجسد (substance → metal, lit. 'body') D
الصمغ] ماء الصمغ (gum arabic → water of gum arabic) D T K L

Loci similes: *Art of Penmanship*, recipe no. 20.

[31] Preparation of an adhesive for gold

Take some fish glue, spread it, put it into water with saffron, put it on a low fire until it has reached a certain consistency and rises from the bottom [in ebullition]. Once it has dried, moisten it with your saliva, spread the glue on top of the gold; once it has dried, burnish it with onyx or with hematite and it will be beautiful.

صفة لصاق الذهب

تأخذ غراء السمك ينشر وتجعله في الماء ويكون بالزعفران ويرفع على نار لينة حتى يأخذ له قوام ويمشي من تحت فاذا جف تنديه بريقك وتلصق من فوقه الذهب فاذا جف فالصقله بالجزع او بحجر الصرف فانه يحسن

فالصقله بالجزع او بحجر الصرف] (burnish it with onyx or with hematite) *om.* T K; in Ba the title is matched with the following recipe too.
وتلصق من فوقه الذهب فاذا جف] (moisten it with your saliva, spread the glue on top of the gold) *om.* D.

Loci similes: Ibn Bādīs, p. 85

[32] Another one like this

Take some white gum ammoniac, dilute it with sublimated wine vinegar, leave it for one hour until they melt; then write whatever you wish with it, on a bow or on a book, and glue the gold or silver leaves on top of it: this will be a good [alternative] to fish glue

اخر مثله

تأخذ الكلخ الابيض ويحل بخل الخمر المصاعد وتتركه ساعة حتى يدوبه ثم اكتب به ما شئت على قوس او كتاب والصق عليه اوراق الذهب او الفضة فانه جيد عن غراء السمك

الكلخ الابيض] الكلخ الابيض وهو الوشق (white gum ammoniac → white gum ammoniac that is [also called] *al-waššaq*) *add.* L; *al-kalḫ* may mean 'giant fennel,' while *al-kalaḫ* can be interpreted as 'gum ammoniac,' a resin that is extracted from the family of plants to which the giant fennel belongs (*Apiaceae*). In L, the copyist felt the need to provide a more common synonym for the gum ammoniac, i.e. *al-waššaq*.
المصاعد] (sublimated) *om.* B
على قوس او كتاب] على اي شيء اردت (on a bow or on a book → on anything you wish) B; على قوس او سيف كتابة (→ on a bow or a sword with an inscription) K T La; considering this last variant, this might be a recipe for inscribing weapons and possibly other objects.
والصق] والقي (glue → place down) T; واطبق (→ superpose) L

[33] Dyeing of the leaves[22]

If you want this, take the leaves you want, soak them in water with alum (*mušabbab*), throw logwood water, or saffron water or indigo flower water, or whatever water you prefer onto this; then spread [the leaves] in the shade on a thick Persian cane until they have dried; when it has dried, burnish it and write on it whatever you want and it will be good.

صباغ الورق

ان اردت ذلك فخذ ما شئت من الورق
وبله في ماء مشبب والقه على ماء البقم
او في ماء الزعفران او في ماء زهر النيلة
الهندية او ما اخترت ثم انشره على قصبة
فارسي غليظة في الظل حتى يجف فاذا
جف اصقله واكتب عليه ما شئت فانه
مليح

ماء مشبب] ما شئت (water with alum → whatever you want)T K L C; the passive participle *mušabbab* is not attested in dictionaries, but I would opt for reading it as a technical 'neologism' that describes a solution of alum in water, considering also that alum has been used as a fixative in dyeing processes already in premodern times.
والقه] وعد القه (throw → dipping it)T K L C

Loci similes: see *Cum Grano Salis*, p. 333 (no. 130.); al-Iskandarī, pp. 182–184 (nos. 210–217).

[34] Cutting the gold leaves

Take a piece of white skin, sew it into the shape of a pillow, stuff it with cotton, with the point of a knife take a gold leaf, spread it onto the pillow, cut the size you need with the knife. Then, take a piece [of gold] and moisten it with your saliva — but only lightly — and apply it onto the sheet, this may be lifted with a cotton cloth; apply fish glue or gum ammoniac on top of it, smooth it with a dry cotton cloth, let it dry, burnish it and this will be amazing. The cutting of silver is done in the same way, and so be aware of this.

قطع اوراق الذهب

يؤخذ قطعة جلدة حور تخيطها شبه
المخدة وتحشي قطن وتأخذ ورقة الذهب
بطرف سكين وتحل على المخدة وتقطع
منها بالسكين على قدر حاجتك ثم تأخذ
قطعة وتبلها بريقك بل خفيف واعملها
على الورقة فانها تشال في القطنة اعملها
على الغراء او الكلخ ودكدكها بالقطنة
الناشفة وخليها تجف واصقلها تجئ غاية
وكذلك تفصيل الفضة فاعلمه

Loci similes: see Fani, p. 77

22 For other procedures to obtain coloured leaves, see Raggetti, *Cum Grano Salis*, p. 333 (No. 130).

2.2 *Invisible and Wondrous Writings*

The 18th chapter of the *ʿUyūn al-ḥaqāʾiq*, among other things, deals with the preparation of invisible inks, whose impression on paper requires a specific trick or stratagem (*ḥīla*) to become visible. The different procedures are defined as different kinds of writings (*kitāba*), which focuses attention on the result rather than on the writing medium. The last preparation is not an invisible ink, but a stimulant for hair growth and it is used to write on the body and to produce an inscription made of hair on the skin.

Chapter eighteen on the deceptions of the 'astrologers' and the manner of their operations

الباب الثامن عشر في غدائر المنجمين وكيفية اعمالها

غدائر المنجمين] اللعب بغرائب المعزمين (deceptions of the 'astrologers' → playful tricks with strange and curious things of the conjurers) B; عزائم المنجمين (→ incantations of the 'astrologers') D; غرائب المنجمين (→ strange and curious things of the astrologers) T K C

في غدائر المنجمين وكيفية اعمالها] (on the deceptions of the 'astrologers' and the manner of their operations) *om.* La

[35] Another trick, concerning [different] kinds of writing

حيلة اخرى في انواع الكتابة

If one writes with it on a plank of wood (?), it will not appear until quenched coal is poured onto it, and then the writing will appear black, as if it were written with [carbon] ink (*midād*).

من كتب بها على ساعدة لا يبان حتى يدر عليها الفحم المطفئ فتظهر الكتابة سوداء كانها بالمداد

It is said that this is the jinns' way of writing (*ḫaṭṭ al-jinn*) and the writing on a piece of wood is produced only by filtering the water.

فيقول ان هذه الكتابة خط الجن والكتابة على ساعدة اما ان يكون باراقة الماء

المطفئ] المسحوق (quenched → ground) T K

ساعدة] ساعة P, the copyist might have read 'in the very moment, on the spot' in this instance, while he reads *sāʿida* in the following sentence. The word *sāʿida* seems to indicate the piece of wood that holds the pulley (see Lane's Lexicon).

الفحم المطفئ] (quenched coal) *om.* B; الفحم المسحوق (→ ground coal) T K

[36] Another [way of] writing
If you write it during the day, you will not see it, while it will appear during the night, as if it were written with gold.
When you want to obtain this, take the gall of a cheetah, the gall of a black dog, and the gall of a hawk; mix them, write with them on thick parchment, and then you will see it during the night as if it were gold, you should know this.

كتابة اخرى
اذ كتبتها بالنهار لا ترى وفي الليل تبان كانها قد كتبت بالذهب

اذا اردت ذلك فلتأخذ مرارة نمر ومرارة كلب اسود ومرارة بازي تخلطها وتكتب بهم في رق غليظ فانه يراها بالليل كانها بالذهب فاعلم

بالذهب] مثل الذهب الابريز (with gold → like red gold) B
تخلطها] (mix them) *om.* D
في رق غليظ] في ورق غليظ بقلم غليظ (on thick parchment → on a thick sheet [of paper] with a thick pen) B; في ورق بقلم غليظ (→ on a sheet [of paper] with a thick pen) D; في رق بقلم غليظ (→ on parchment with a thick pen) T K C; بقلم غليظ (→ with a thick pen) L
بالليل كانها بالذهب] يكون ذلك (during the night as if it were gold → it will become this) B; في الليل كانه الذهب الابريز (→ in the night as if it were pure gold) D
Loci similes: in terms of different kinds of gall, several recipe describe gall-based inks that become visible and shine like gold at night, see Fani, pp. 48, 49 and 153; al-Rāzī, pp. 226–227 (nos. 40 and 44); al-Qalalūsī, p. 36; al-Iskandarī, p. 180 (no. 208)

[37] Another [way of] writing
Write on the sheet and it will not appear until you place it near to the fire, and then it will appear.
This is that you write using onion water.

كتابة اخرى
تكتب على الورق فلا تبان حتى تقربها من النار فتبان
وهو ان تكتبها بماء البصل

بماء البصل] وتقربها الى النار فانها تبان كتابته حمراء فاعلم ذلك (with onion water → and you place it near to the fire, then its writing will appear red, you should know this) *add.* L

Loci similes: Fani, pp. 47 and 152; al-Qalalūsī, p. 36.

[38] Another [way of] writing that does not appear without a stratagem

Take some sour milk and sal ammoniac, write a message with it, send it to whomever you want, and nothing will appear in it.

When you place it near to the fire, the writing will appear.

كتابة اخرى لا تبان الا بالحيلة

تأخذ لبن حليب ونشادر وتكتب به في كتاب وترسله الى من تريد فانه لا يبان فيه شيء

فمتى ما تقرب من النار ظهرت الكتابة

لا تبان الا بالحيلة] (that does not appear without a stratagem) *om.* B T L

لبن حليب] لبن طيب (sour milk → good milk) T; حليب (→ milk) L

ونشادر] والعقاب (sal ammoniac → the eagle) B, this is a possible code name (*Deckname*) for sal ammoniac, see Siggel, *Decknamen*, pp. 18 and 45.

وتكتب...شيء] وتنظره لمن اردت (write [...] appear in it → and show it to whoever you want) B

Loci similes: al-Iskandarī, p. 180 (no. 206), the recipe here suggests to pour water on the writing.

[39] Another [way of] writing

Write with vitriol water on any leaf you want; when you want it to appear, throw it into gall nut water, and then a black writing will appear.

كتابة اخرى

تكتب بماء الزاج على ما اردت من الاوراق فاذا اردت اظهاره فالقها في ماء العفص فانها تظهر كتابة سوداء

في ماء العفص] في ماء مذاب فيه عفص منقوع فانها تظهر سوداء (in gall nut water → in the water that has been mixed with macerated gall nuts) C, this variant gives a more precise indication for understanding what, more in general, 'the water of (any ingredient)' might be.

Loci similes: see recipe no. 16 in Ch. 23 of the *ʿUyūn al-ḥaqāʾiq*; Fani, p. 152; al-Qalalūsī, p. 36; *Cum Grano Salis*, p. 331 (no. 120), al-Zarḫūrī mentions only 'white water' that the editor interprets as 'transparent water'; 'white vitriol water,' the editor identifies it with 'the sulphurs of the spear makers,' this recipes seems to mention two different vitriols, one to write the other to make the writing appear, see al-Rāzī, p. 225 (no. 26); al-Iskandarī, p. 179 (no. 205). Often, the process is inverted: one has to write with gall nut water and make it appear with vitriol.

[40] Another [way of] writing

If you wish to write on a red leaf or on a blue one with a writing that appears as if it were made with silver, then take some quicksilver, pour some tin over it, and calcinate it with this; its blackness will be extracted from it; make it into a powder. Add gum arabic water to it and write with it what you wish.

When the writing has dried, polish it with onyx, and then it will appear as if it had been written with silver.

كتابة اخرى

اذا اردت ان تكتب على ورقة حمراء او زرقاء كتابة تظهر كانها كتابة بالفضة فتأخذ من الزيبق وتلقي على المشترى وتكلس به ويخرج سوادة عنه واجعل تربة وتسقيها بماء الصمغ واكتب به ما شئت

فاذا جفت الكتابة فاصقلها بالجزع فانها تظهر كانها قد كتبت بالفضة

حمراء او زرقاء] حمراء او صفراء او زرقاء (on a red leaf or on a blue one → on a red, yellow, or blue leaf) *add.* La.

المشترى] القصدير (Jupiter → tin) B T K L C, Jupiter' is a common code name (*Deckname*) for tin, see Siggel, *Decknamen*, pp. 18 and 45.

واجعل تربة] واجعله توتية (make it into a powder → make it into zinc) B; برده (→ file it) C

بالجزع] بحجر الصقل (with onyx → with a burnishing stone) C

[41] Another [way of] writing

Write it on a leaf not treated with starch and it will not appear; when you throw it in water, then a white writing will appear.

When you wish this, take some Yemeni alum dissolved in wine vinegar, and write with it what you want.

When it has dried, throw it in water and what we have mentioned will appear from it.

كتابة اخرى

تكتبها على الورق غير المنشاء فلا تبان فاذا القيتها في الماء فانها تبان كتابة بيضاء

فاذا اردت ذلك فخذ الشب اليماني المحلول بخل الخمر واكتب به ما شئت

فاذا جف تلقي في الماء فانه يبان منه ما ذكرناه

المنشاء] مشى (treated with starch → walking, going?) B, perhaps the simplification of a technical term.

بخل الخمر] بخل خمر مقطر (with wine vinegar → with distilled wine vinegar) B; بالخل المقطر (→ with distilled vinegar) D T K C

[42] Another [way of] writing
Take some black cumin, egg yolk, and the peelings of colocynth roots fried in good oil; when you write with it on any area of the body, then hair will grow on the spot, so you should know this and hide it from the ignorant ones.

كتابة اخرى
تأخذ كمون اسود وصفار البيض وقشور عروق حنظل المقلية بالزيت الطيب اذا كتبت به على مكان في جسد فانه ينبت مكانه الشعر فاعلم واكتم عن الجهال

وصفار البيض] ودهن صفار البيض (egg yolk → oil of egg yolk) *add.* B D T L K C La
المقلية] المغلي (fried → boiled) B
واكتم عن الجهال] (hide it from the ignorant ones) *om.* B T K La; فانه من الاسرار واكتمه (→ in fact this belongs to the secrets and you must hide it) L

3 Concluding Remarks

The 13th century was a time of literary interest in the explanation or unveiling of technical tricks, frauds, and dupes. Al-ʿIrāqī's *ʿUyūn al-ḥaqāʾiq* represents an interesting case in the genre and collects many different streams of tradition: pseudo-Platonic magic, Galenic medicine, occult properties, talismans, sleight of hand, and different crafts. The preparation of metallic inks and invisible writing media can be accounted for in this last component.

Although no direct source is unequivocally identified, the recipes here have many parallels in technical handbooks and texts on ink making. The distribution in two different chapters, however, and the order of the recipes within them seems original, possibly determined by the different kinds of composition and textual genre. The lack of parallel attestations for some recipes may indicate that they might be procedures of al-ʿIrāqī's own invention, or, alternatively, of other sources still to be identified.

The overall structure of the text — introduction, division into 30 chapters — remains constant throughout the tradition, while the contents of the single chapters and their wording are transmitted in a fluid way with a high degree of variance. The distribution of the recipes displayed in Appendix II and the variance highlighted by the edition shows the fundamental importance of a *recensio* that aims at completeness, even more in the case of fluid traditions. Preferring a single witness over a number of others would result in a significant loss of information.

The interdisciplinary collaboration for the study of premodern science and technology requires the support of specific tools. The contribution of the philologist may consist of an edition that highlights the technical aspects of the text — a 'laboratory-edition' meant for interdisciplinary use — and makes technical variants available to a larger readership.

Appendix 1: Descriptive Table of Contents of the *Kitāb ʿuyūn al-ḥaqāʾiq*

Chapter 1: on the *nawāmīs* (lit. 'laws' [of nature]) **and the manner of their operations.**	الباب الاول في النواميس وكيفية اعمالها

Plato is indicated as a source; there are two kinds of *nawāmīs*, a high and a low one. As for the former, God has given it to high-ranking people (*ahl al-darağāt*) who produce wonders, such as making the moon appear during the day and the sun during the night, who can affect lightning, thunder, wind, and the sea, trees and fruits.

Food *nawāmīš*: a small quantity of this food makes one grow a lot.

Pills that allow someone to live for one month without drinking or sleeping; make a camel able to travel for one month (attributed to Aristotle); a preparation for walking on water (attributed to al-Rāzī), sometimes, the preparations may include the writing of magical signs and names.

'The people of the alteration' (*Ahl al-taṣrīf*): open doors; disappear; levitate; the occult properties of letters (*ḫawāṣṣ al-ḥurūf*).

Chapter 2: on incendiary preparations (*maḥārīq*) and the manner of their operations.	الباب الثاني في المحاريق وكيفية اعمالها

Incendiary preparations that, once kindled, give the impression that the house or the place is shining like gold or some other colour, or that the angels are descending in the house, or that the light gathers around someone at night; for eating all kinds of fruits and vegetables out of season.

Chapter 3: on fumigations (*al-daḫan*) and the manner of their operations.	الباب الثالث في الدخن وكيفية اعمالها

Fumigations, often attributed to Plato, which create the effect that darkness has fallen on earth; that make an army appear to the eye of the beholder; that make appear stars and huge birds in the sky; that will make a tree bend towards the person sitting underneath it (one specific for the palm tree); that create the impression among a group of people that they have turned into elephants or large beasts; that summon mice; that create the impression that a crocodile is coming out of the water; that summon jinns and evil spirits; and powerful fumigations that wise men from India and from Babylon used to affect the luminaires and atmospheric phenomena.

Chapter 4: on fermentations (*al-taʿāfīn*) and the manner of their operations. الباب الرابع في التعافين وكيفية اعمالها

After a concise theoretical explanation of spontaneous generation, different operations that involve the putrefaction of animal substances that produce strange creatures whose properties are to be exploited.

Chapter 5: on soporifics (*al-marāqid*) and the manner of their operations. الباب الخامس في المراقد وكيفية اعمالها

Different preparations (potions, lanterns, apples, drinks) with an immediate soporiferous effect; at the end of the chapter there is a description of two powerful poisons, one of them attributed to Aristotle (*iksīr li-halāk*, 'elixir of annihilation').

Chapter 6: on astrological incantations (*al-nāringǧāt*) and the manner of their operations. الباب السادس في النارنجات وكيفية اعمالها

Different potions and preparations to provoke love and hatred (attributed to Hermes and Sāsān); to make clothes become infested with lice; to extract a tooth; to provoke a disease; to cause tears.

Chapter 7: on concealments (*al-iḫfāt*) and the manner of their operations. الباب السابع في الاخفآت وكيفية اعمالها

Different procedures for becoming invisible during the day and at night, usually implying the use of specific animals (cat, hoopoe, frog)

Chapter 8: on illusions and tricks (*al-dakk wa-l-ḥiyal*) and the manner of their operations. الباب الثامن في الدك والحيل وكيفية اعمالها

Procedures involving ritual prescriptions (such as sitting in a hoopoe cage for 40 days) and the use of secret names and magical figures in order to obtain various things, or to create an illusion, for instance how appear to levitate in the sky or to walk in the fire without burning.

Chapter 9: on cultivations (*zirāʿāt*) and the manner of their operations.

الباب التاسع في الزراعات وكيفية اعمالها

Procedure for the instant germination of seeds and plants.

Chapter 10: on amusing tricks (*laʿb*) with eggs and the manner of their operations.

الباب العاشر في اللعب بالبيض وكيفية اعمالها

A number of amusing tricks performed with eggs, for instance to peel it and make some writing appear underneath the shell, or to give the impression that the egg is flying.

Chapter 11: on amusing tricks with bottles (*qanānin*) and the manner of their operations.

الباب الحادي عشر في اللعب بالقناني وكيفية اعمالها

A number of amusing tricks performed with bottles, for instance a bottle whose opening is lit like a candle, or a bottle whose contents can boil without any fire.

Chapter 12: on amusing tricks with seals (*ḫawātīm*) and the manner of their operations.

الباب الثاني عشر في اللعب بالخواتيم وكيفية اعمالها

A number of amusing tricks performed with seals, for instance a seal that moves on a hard surface like a tile or a stone.

Chapter 13: on amusing tricks with effigies (*tamāṯīl*) and the manner of their operations.

الباب الثالث عشر في اللعب بالتماثيل وكيفية اعمالها

A number of amusing tricks performed using wax figurines with animal or human shape, which, for instance, do not melt in the fire or are able to keep flies away, or that may take on a specific colour when exposed to heat.

Chapter 14: on amusing tricks with arrows (*aqdāḥ*) and the manner of their operations.

الباب الرابع عشر في اللعب بالاقداح وكيفية اعمالها

A number of amusing tricks performed with arrows, for instance an arrow that bends without breaking, or arrows filled with different liquids.

Chapter 15: on amusing tricks related to slaughterings (*ḏabāʾiḥ*) and the manner of their operations.

الباب الخامس عشر في اللعب بالذبائح وكيفية اعمالها

A number of amusing tricks performed with wax figurines that twitch or bleed when cut.

Chapter 16: on amusing tricks with fire and the manner of their operations.

الباب السادس عشر في اللعب بالنار وكيفية اعمالها

A number of amusing tricks performed with fire in order, for instance, to hold it in one's mouth or to set clothes ablaze without burning them, prepare incendiary figurines, or enter a burning furnace.

Chapter 17: on amusing tricks with lanterns (*suruǧ*) and the manner of their operations.

الباب السابع عشر في اللعب بالسرج وكيفية اعمالها

A number of amusing tricks performed with lamps and lanterns that, for instance, give the impression that the house is full of snakes or scorpions, or make green birds appear, or a naked woman who starts dancing, or that have the power to make people appear like statues or like all sorts of animals to others in the same room.

Chapter 18: on the deceptions of the 'astrologers' and the manner of their operations.[23]

الباب الثامن عشر في غدائر المنجمين وكيفية اعمالها

Different procedures focused on the use of inscriptions on paper or papyrus, wax figurines and invisible inks; the purpose is achieved also thanks to the recitation of pious formulae. Written names can be used, for instance to get known thieves out of hiding, invisible inks and wondrous ways of writing.

Chapter 19: on tricks of dexterity by those 'who play tricks with the stick' (*tanābīl al-muǧarridīn*) and the manner of their operations.

الباب التاسع عشر في تنابيل المجردين وكيفية اعمالها

---

23 See Höglmeier, *Al-Ǧawbarī* (cit. note 2), p. 214; for the variant readings attested in this chapter heading, see the edition in this chapter.

Tricks by this category of tricksters, whose aim is ultimately to gain from every situation, to dupe naïve bystanders by pretending, for instance, to be able to drink normal water and spit rose water (by means of a compress hidden under the tongue); there are many techniques for approaching potential victims and these involve expressing words of appreciation for a seal ring, a warning against scorpions and snakes, a round of three-shell game, etc.

Chapter 20: on those who play tricks with slips of paper (*qays al-mušarmiṭīn*) and the manner of their operations.	الباب العشرون في قيس المشرمطين وكيفية اعمالها

Tricks to make a certain name appear on paper, sometimes based on the knowledge of secret written names or signs; to prepare a shirt inscribed with amulets that protects against any wound, etc.

Chapter 21: on the conditions of the cattle (*aḥwāl al-dakāšira*) and the manner of their operations.	الباب الحادي والعشرون في احوال الدكاشرة وكيفية اعمالها

Tricks played on animals (for instance to provoke an epileptic seizure) to lower their price; or the use of animal ingredients, and animal-based hair dyes.

Chapter 22: on remedies for hunting different kinds of animals (*adwiya ṣayd ağnās al-ḥayawān*) and the manner of their operations.	الباب الثاني والعشرون في ادوية صيد اجناس الحيوان وكيفية اعمالها

Remedies and preparations to propitiate and ease the hunting of wild animals, such as the lion, and the crocodile; for fishing; for the preparation of deadly poisons.

Chapter 23: on kinds of metallic inks and dyes (*al-liyaq wa-l-asbaġ*) and the manner of their operations.	الباب الثالث والعشرون في انواع الليق والاصباغ وكيفية اعمالها

Recipes for coloured metallic preparations that can be used to write or paint.

Chapter 24: on simple drugs (*al-adwiya al-mufrada*) and the manner of their operations.	الباب الرابع والعشرون في الادوية المفردة وكيفية اعمالها

Useful properties derived from the practical experiences (*taǧārib*) of ancient men, grouped by the purpose they serve or by the kind of preparation (pills, powder, ointment, etc.)

Chapter 25: on dyestuffs and dyes (*al-ḫiḍābāt wa-l-ṣibāġāt*) and the manner of their operations.	الباب الخامس والعشرون في الخضابات وكيفية اعمالها

Recipes to prepare dyes of different colours (black, gold, green, blue) for different materials and black dyes for the hair.

Chapter 26: on simple artifices (*al-malāʿīb al-mufrada*) and the manner of their operations.	الباب السادس والعشرون في الملاعب المفردة وكيفية اعمالها

Procedure to produce an illusion using an image painted on a wall; to make a severed head of an animal emit a cry; to gather birds in a certain place; to make a dog dance, etc.

Chapter 27: on the occult properties of metals and stones (*ḫawāṣṣ maʿādin wa-l-aḥǧār*) and the manner of their operations.	الباب السابع والعشرون في خواص المعادن والاحجار وكيفية اعمالها

Association of stones with the seven planets and explanation of their alchemical and medical occult properties; Hermes is presented as the source for this material.

Chapter 28: on the occult properties of plants (*ḫawāṣṣ al-nabāt*) and the manner of their operations.	الباب الثامن والعشرون في خواص النبات وكيفية اعمالها

Various properties of plants applied to practical jokes and healing.

Chapter 29: on the occult properties of animals (*ḫawāṣṣ al-ḥayawān*) and the manner of their operations.	الباب التاسع والعشرون في خواص الحيوان وكيفية اعمالها

On the useful and occult properties of animal parts, a selection that does not seem to be arranged in a particular order.

Chapter 30: on the manner of the composition of man (*tarkīb al-insān*) and the peculiarities that come with the passing of time (turning of the seasons).

الباب الثلاثون في كيفية تركيب الانسان وما يختص به على ممر الزمان

The contents are presented as al-ʿIrāqī's, who states the superiority of men over all the other beings and the correspondence between macrocosmos and microcosmos, the special regimen required by the different seasons based on the theory of the four qualities.

Appendix 11: Synoptic Table of the Recipes as Attested in the Different Manuscript Witnesses

TABLE 8.1 Metallic inks (*liyaq*)

Chapter 23		P ff. 94r–99r	B ff. 52v–54r	D ff. 52r–54r	T pp. 84–82	K ff. 52r–53v	L ff. 74v–76r	C pp. 38–40	La ff. 23r–24v
1.	Preparation of gum arabic	✓	✓	✓	✓	✓	✓	✓	✓
2.	Cinnabar ink	✓	✓	✓	✓	✓	✓	✓	✓
3.	Arsenic ink	✓	✓	✓	✓	✓	✓	✓	✓
4.	Another [arsenic] ink	✓	✓	✓	✓	✓	✓	✗	✓
5.	Green ink	✓	✓	✓	✓	✓ + Gold water	✓	✓	✓
6.	Verdigris ink	✓	✓	✓	✓	✓	✓	✓	✓
7.	Pistachio ink	✓	✗	✓	✓	✓	✗	✓	✗
8.	White ink	✓	✓	✓	✓	✓	✓	✓	✗
9.	Lapis lazuli ink	✓	✓	✓	✓	✓	✓	✓	✗
10.	Wine ink	✓	✓	✓	✓	✓	✓	✗	✗
11.	Turtle-dove ink	✓	✓	✓	✓	✓	✓	✗	✗

TABLE 8.1 Metallic inks (*liyaq*) (*cont.*)

Chapter 23		P ff. 94r–99r	B ff. 52v–54r	D ff. 52r–54r	T pp. 84–82	K ff. 52r–53v	L ff. 74v–76r	C pp. 38–40	La ff. 23r–24v
12.	Rosy ink	✓	✗	✗	✗	✗	✗	✗	✗
13.	Orange ink	✓	✗	✗	✗	✗	✗	✗	✗
14.	Honey ink	✓	✗	✗	✗	✗	✗	✗	✗
15.	Silver ink	✓	✗	✗	✗	✗	✗	✗	✗
16.	Black vitriolic ink	✓	✗	✗	✗	✗	✗	✗	✗
17.	Golden ink	✓	✗	✗	✗	✗	✗	✗	✗
18.	Ashen ink	✓	✗	✗	✗	✗	✗	✗	✗
19.	Violet ink	✓	✗	✗	✗	✗	✗	✗	✗
20.	Turquoise ink	✓	✗	✗	✗	✗	✗	✗	✗
21.	Peony ink	✓	✗	✗	✗	✗	✗	✗	✗
22.	Clay ink	✓	✗	✗	✗	✗	✗	✗	✗
23.	Lentil ink	✓	✗	✗	✗	✗	✗	✗	✗
24.	Abbasid ink	✓	✗	✗	✗	✗	✗	✗	✗
25.	Ivory ink	✓	✗	✗	✗	✗	✗	✗	✗
26.	Golden ink	✓	✗	✗	✗	✗	✗	✗	✗
27.	Preparation for golden, silver, copper, and lead ink	✓	✗	✗	✗	✗	✗	✗	✗
28.	Wood ink	✓	✓	✓	✓	✓	✓	✗	✗
29.	Section on the melting of gold	✓	✓	✓	✓	✓	✓	✓	✗
30.	Melting of all the metals	✓	✗	✓	✓	✓	✓	✗	✗
31.	Preparation of an adhesive for gold	✓	✗	✓	✓	✓	✓	✗	✗

Chapter 23	P ff. 94r–99r	B ff. 52v–54r	D ff. 52r–54r	T pp. 84–82	K ff. 52r–53v	L ff. 74v–76r	C pp. 38–40	La ff. 23r–24v
32. Another one like this	✓	✓	✓	✓	✓	✓	✗	✓
33. Dyeing of the leaves	✓	✓	✓	✓	✓	✓	✓	✗
34. Cutting of the gold leaves	✓	✗	✗	✗	✗	✗	✗	✗

TABLE 8.2 Invisible writing (*kitāba*)

Chapter 18	P ff. 63r–72r	B ff. 40v–42v	D ff. 41r–41v	T pp. 58–66	K ff. 37v–42v	L ff. 66v–68r	C pp. 32–34	La f. 18r
35. [Black coal writing]	✓	✓	✓	✓	✓	✗	✗	✗
36. [Night writing]	✓	✓	✓	✓	✓	✓	✓	✗
37. [Onion writing]	✓	✓	✓	✓	✓	✓	✗	✓
38. [Sal ammoniac writing]	✓	✓	✓	✓	✓	✓	✗	✗
39. [Vitriol writing]	✓	✓	✓	✓	✓	✓	✓	✗
40. [Silver writing on coloured inks]	✓	✓	✓	✓	✓	✗	✓	✓
41. [Yemeni alum writing]	✓	✓	✓	✓	✓	✗	✓	✗
42. [Hair growth writing]	✓	✓	✓	✓	✓	✓	✗	✓

Index of Manuscripts

Index of Authors

Index of Sources

Index of Technical Terms